SERIES

ENTOMOLOGICA

EDITOR

E. SCHIMITSCHEK

Göttingen

VOLUMEN 3

Springer-Science+Business Media, B.V.

A REVISION OF THE FAR EAST ASIAN APHIDIIDAE (HYMENOPTERA)

by

PETER STARY

Prague

&

EVERT I. SCHLINGER

Riverside, Cal.

Springer-Science+Business Media, B.V.

ISBN 978-94-017-5859-8   ISBN 978-94-017-6337-0 (eBook)
DOI 10.1007/978-94-017-6337-0
Copyright Springer Science+Business Media Dordrecht 1967
Originally published by Dr W. Junk Publisher The Hague in 1967.
Softcover reprint of the hardcover 1st edition 1967

CONTENTS

INTRODUCTION

In 1961, the junior author enjoyed a four-months expedition to several Far East Asian countries, in search of possible natural enemies of aphids to import into California, for the purpose of controlling biologically certain species of pest aphids. A considerable effort also was made at that time to collect, rear and host-associate any type of aphid parasite encountered. At the same time the senior author expressed his interest in studying these parasites from East Asia, hence this joint work was undertaken. This is a report of the taxonomic and some bio-ecological data ascertained from the reared specimens and a few other specimens gathered for study from sources in Czechoslovakia, the USSR and the United States National Museum.

The species included here were collected in Hong Kong, Taiwan, Japan (Kyushu, Honshu and Hokkaido Islands), South Korea and eastern USSR and are represented by host-associated specimens.

Altogether 62 species in 18 genera and subgenera are described and of these taxa, 3 genera, 1 subgenus and 14 species are described as new. Holotype specimens will be deposited in the U.S. National Museum, Washington, D.C., and paratypes at both of the authors. institutions and with Dr. M. J. P. MACKAUER* in Canada as available.

ACKNOWLEDGEMENTS

In a study of this kind the help and interest of numerous persons has been invaluable aid to the authors and we gratefully acknowledge them as follows:

Hong Kong – Mr. S. K. CHENG, Chisun Horticultural Co., Shatin, H.K.
  Mr. P. Y. SO, Agriculture and Forestry Dept., Kowloon

Taiwan – Mr. Charles CHIA-CHU TAO and Mrs. S. CHEN-CHIU of the Taiwan Agricultural Institute, Taipei.
  Dr. TINGWEI LEW, Joint Commission on Rural Reconstruction, Taipei.
  Mr. F. T. YAO, Director of Horticulture Experiment Station, Hualien.

Japan – Dr. KEIZO YASUMATSU, Dr. Yoshihiro HIRASHIMA and
  Dr. HIDAKA from Kyushu University, Fukuoka.

---

* Entomological Institute for Biological Control, Belleville, Ontario; formerly Zoological Institute, The University, Frankfurt M., Germany.

Dr. Akira NAGATOMI, University of Kagoshima.

The late Dr. R. TAKAHASHI, Osaka.

Dr. M. SASAGAWA, Kyoto University.

The late Dr. S. KATO, Tokyo.

Mr. Kaku SATO, Imperial Plant Quarantine Service, Tokyo.

Dr. Hitoshi HASEGAWA, National Institute of Agricultural Science, Tokyo.

Dr. Chihisa WATANABE, Hokkaido University, Sapporo.

Dr. Motonori INOUYE, Forestry Experiment Station, Sapporo.

South Korea – Dr. WOON HAH PAIK, Seoul National University, Suwon.

United States – Dr. C. F. W. MUESEBECK, United States National Museum, Washington, D.C.

USSR – Dr. V. TOBIAS, Zoological Institute, Academy of Sciences of USSR, Leningrad.

Many other persons not mentioned from these areas have also lent assistance to our study and to each of them we also express our gratitude.

Authors' addresses:

Dr. P. Starý, Institute of Entomology, Czechoslovak Academy of Science, Prague, Czechoslovakia.

Dr. E. I. Schlinger, Department of Entomology, University of California, Riverside, Calif., U.S.A.

BIOLOGICAL CONTROL

Even though it is not the intent of this paper to discuss the biological control of aphids, it seems desirable to briefly note recent work in this field of interest and comment on the program that took the junior author to East Asia.

Several aphid species have in the past 10 years caused such economic problems to crops in California that the biological aspect of their control was begun. The first of these was the spotted alfalfa aphid, *Therioaphis maculata* (BUCKTON)*, a serious pest of *Medicago sativa* which is now considered a very minor pest due primarily to a successful biological control program as discussed and shown by VAN DEN BOSCH (1957), VAN DEN BOSCH et al. (1957, 1959 a and 1959 b). The phenological study of the three imported parasites of this aphid has only recently been completed (VAN DEN BOSCH, et al. 1964). Another important aphid pest under a considerable degree of biological control in California is *Acyrthosiphon pisum* (HARRIS) (=*Macrosiphum pisi*). Until the importation of the parasite *Aphidius smithi* SHARMA & SUBBA RAO, from India to California in 1957 – 58, this aphid was also a serious pest of alfalfa and peas. Biological control of this aphid was remarkable in a very short time in southern California (HAGEN & SCHLINGER, 1960) and has continued to spread into wider areas each year. Further research by VAN DEN BOSCH, HAGEN and SCHLINGER is in press, and one important aspect of this study (VAN DEN BOSCH & SCHLINGER, 1964) was a clear demonstration of the density-dependent action of the host-parasite relationship.

A third aphid, *Chromaphis juglandicola* (KALTENBACH), a pest of *Juglans*, has recently been under a biological control study, and further research on the success of its imported European parasite, *Trioxys pallidus* HALIDAY, is continuing. SCHLINGER et al. (1960) and VAN DEN BOSCH et al. (1962) have published preliminary observations on the results.

Two aphid pest species, *Aphis spiraecola* PATCH and *A. gossypii* GLOVER, were the primary reasons for the junior author's visit to East Asia in 1961. A number of parasites were located, reared and sent to California for mass rearing and release, but to date none of those obtained seem to be established.

* According to recent correspondence with Dr. R. VAN DEN BOSCH, the name of this aphid is being changed to *Therioaphis trifolii* (MONELL) *sensu lato*.

2

The reasons for this are still somewhat unknown. Attempts are being made by Dr. VAN DEN BOSCH to obtain other parasites of these aphids.

Aphis spiraecola is an intermittent, but often serious pest of *Citrus, Spiraea,* and a number of ornamental plants throughout North America and in other regions. Since no "good" records of parasites of this aphid were known in North America, since parasites in Europe were also ineffective (personal communication from R. VAN DEN BOSCH), and because the aphids primary host *(Spiraea)* is known to be represented by many species in East Asia, it was believed that somewhere in East Asia valuable parasites could be located. During the junior author's travel in East Asia, 6 aphidiid parasites in 5 genera were found in Taiwan, Japan and South Korea. At least one species of aphelinid *(Aphelinus)* was also found, as well as numerous predators. None of these parasites when imported were successful, and few if any of them were found to be host-specific. It is hoped, however, that somewhere in China or the Himalayan region parasites that are more efficient might some day be located.

Parasites of *Aphis gossypii* (6 species in 4 genera) were likewise located in East Asia and imported to California. None of these have been successful to date.

Biological control studies of aphids are only beginning and we hope for success in the future.

BIOLOGY AND HOST RELATIONSHIPS

It is well known that aphidiids are all primary, solitary, internal parasites of aphids. Many excellent papers on biology, ecology and host relationships of the Aphidiidae have appeared within the past few years, and even though it is not our intent to review or discuss these subjects to any great degree, it seems important to mention some of them, as a few do pertain to species in Far East Asia.

Articles which deal mainly with biology, morphology and bio-ecology are: MILLAN (1956) on *Aphidius platensis* BRÈTHES; GEORGE (1957) on several parasites; SCHLINGER (1960) on several parasites; SCHLINGER & HALL (1959, 1960 b, 1961) on *Praon palitans* MUESEBECK and *Trioxys utilis* MUESEBECK; HALL et al. (1962) on "sibling" *Trioxys utilis* MUESEBECK and *T. pallidus* HALIDAY; SUBBA RAO & SHARMA (1962) on *Trioxys indicus* SUBBA RAO and SHARMA; STARÝ (1962 d) on *Aphidius ervi* HALIDAY; STARÝ (1962 c) on *Ephedrus persicae* FROGGATT *(as E. pulchellus)*; WIACKOWSKI (1962) on *Aphidius smithi* SHARMA and SUBBA RAO; SCHLINGER & MACKAUER (1963) on *Aphidius matricariae* HALIDAY; TREMBLAY (1964) on *Lysiphlebus fabarum* (MARSHALL); and MESSENGER & FORCE (1963) on *Praon palitans* MUESEBECK.

Only quite recently have any real attempts been made to study host selection and host relationships. SCHLINGER & HALL (1960 a) attempted to clarify the specificity of a few parasite species. More detailed papers by MACKAUER (1961 c, 1962 c and 1963) and STARÝ (1963 a, b, c, d and 1964) have shown that all degrees of host selection from monophagy to polyphagy occur, as well as various degrees of restriction to higher taxonomic levels of the host. It is clear, however, that in the fields of both biology and host selection more detailed studies are needed before reliable generalizations can be made.

ZOOGEOGRAPHY

The Far East Asian aphidiid fauna as now known consists of 62 species representing 18 genera and subgenera. Table 1 lists each of the areas visited for this report, the numbers of recorded and new taxa for each area, and the number of the known taxa for the whole area.

Table 1.	*Comparison of new vs. recorded taxa of Far East Asian Aphidiidae*					
	Recorded		New Records of		Total Known Taxa	
Locality	Genera & Subgenera	Species	Genera & Subgenera	Species	Genera & Subgenera	Species
Hong Kong	3	3	4	6	7	9
Taiwan	7	9	5	12	12	21
South Korea	1	1	8	16	9	17
Japan:						
Kyushu Island	1	1	9	15	10	16
Honshu Island	3	3	6	8	9	11
Hokkaido Island	3	9	3	11	6	20
Totals	10	24	8	38	18	62

ORIENTAL REGION

Hong Kong. There are only 9 species in 7 genera and subgenera now recorded from this small, subtropical, southern tip of China. None of the genera or subgenera are endemic, and only one of the species (*Trioxys (Binodoxys) sinensis* MACKAUER) is not known elsewhere. Other zoogeographic relations are shown by the presence of the rather cosmopolitan *Ephedrus persicae* FROGGATT, the Palearctic-Oriental *Ephedrus plagiator* (NEES) and *Lipolexis gracilis* FÖRSTER. *Lipolexis scutellaris* MACKAUER and *Trioxys (Fissicaudus) confucius* MACKAUER occur in Hong Kong and Taiwan, while *Lysiphlebia rugosa* STARÝ and SCHLINGER is known otherwise only from South Korea. *Praon orientale* STARÝ and SCHLINGER is a typical Far East Asian species, recorded from all the islands and areas visited. *Pauesia unilachni* (GAHAN) is a species now recorded from Taiwan and South Korea, but remains unknown from any of the Japanese islands. Although Hong Kong certainly fits into the "Oriental Region" of WALLACE, the Aphidiidae of Hong Kong show a rather strong faunal relationship to the Palearctic region, particularly continental Far East Asia and Europe.

Taiwan. This island has the largest recorded aphidiid fauna in the Far East Asia. 21 species in 12 genera and subgenera are recorded. One genus, (the monotypic *Archaphidus* STARÝ and SCHLINGER), is endemic. Nine of the 21 species appear to be endemic. These apparent endemics are in the following genera and subgenera: *Trioxys (Trioxys)*; *Trioxys (Binodoxys)*; *Ephedrus*; *Monoctonus*; *Pauesia*; and *Aphidius*. The nearly cosmopolitan species, *Ephedrus persicae* FROGGATT is present, as are also the widespread palearctic-oriental *Ephedrus plagiator* (NEES), *E. lacertosus* (HALIDAY), *Aphidius absinthii* MARSHALL, *A. salicis* HALIDAY and *Lipolexis gracilis* FÖRSTER. Of the 8 species of *Trioxys* (including *Binodoxys*) recorded from Taiwan, all but *T.(B.) ndicus* SUBBA RAO and SHARMA (from India) appear to be endemic. Only a single record of *Praon* is know from Taiwan, a rather strange fact since *Praon* (although not rich in species) was found to be common in all other Far East Asian areas that were visited. Such species as *Aphidius salignae* WATANABE, *A. "gifuensis"* and *Lysiphlebia japonica* (ASHMEAD) show to a faunal relation to Japan, while species like *Pauesia unilachni* (GAHAN), *Aphidius absinthii* MARSHALL, and *A. salicis* HALIDAY show an interesting discontinuous distribution, occurring also in South Korea, Hong Kong and / or only the northern-most Japanese island, Hokkaido. Only 3 of the 21 species in Taiwan are found widely throughout Far East Asia. The fact that Taiwan appears to have a somewhat unique and separately-evolved aphid fauna might indicate that further new aphidiid parasites are awaiting discovery.

Taiwan, even with its rather distinct fauna, has faunal relationships with the Oriental, European, Palearctic and continental Far East Asian areas and to a much lesser degree with the Japanese islands of Kyushu and Honshu. As an example of the latter, the maple aphids (*Periphyllus* spp.) were found to be present throughout the Far East on both *Acer* and *Koelreuteria*, and all over Japan and South Korea these aphids were almost invariably found with an associated parasite, *Aphidius areolatus* ASHMEAD. But during a one month's study, not a single parasite was located in Taiwan attacking the immense populations of *Periphyllus* on *Acer* in the Taipei Botanical Garden and other areas. This parasite apparently has not yet invaded Taiwan from the north, nor has it been able to migrate from China.

Kyushu Island – Japan. This is the southern-most of the large islands of Japan and climatically the warmest. The aphidiid fauna is a zoogeographic mixture. It contains 16 known species in 10 genera and subgenera. There is one endemic genus, the monotypic *Bioxys* STARÝ and SCHLINGER, and there are 4 apparently endemic species, one each in *Paralipsis* FÖRSTER, *Pauesia* QUILIS, *Aphidius* NEES and *Praon* HALIDAY. The non-endemic species show relationships to almost all other areas, such as: *Praon orientale* STARÝ and SCHLINGER, *Aphidius absinthii* MARSHALL, *A. salignae* WATANABE, *A.* "*gifuensis*" group and *Lysiphlebia japonica* (ASHMEAD) which occur in all of Far East Asia and some of them in western Palearctic areas as well; *Lipolexis gracilis* FÖRSTER, which is associated with the Oriental region as well as with western Palearctic areas. Two species show relationships to South Korea and Honshu Island. These are: *Trioxys (Binodoxys) orientalis* STARÝ and SCHLINGER and *Diaeretus leucopterus* (HALIDAY). *Aphidius areolatus* ASHMEAD is known also only from South Korea, Honshu and Hokkaido.

Honshu Island – Japan. This is the largest of the Japanese islands and contains subtropical to near arctic faunal elements. The Aphidiidae of Honshu include 11 known species in 9 genera and subgenera. There are no endemic genera and only 4 possibly endemic species. These are: *Pauesia japonica* (ASHMEAD), *Trioxys (Binodoxys) brunnescens* STARÝ and SCHLINGER, *Praon quadratum* STARÝ and SCHLINGER and *Protaphidius nawaii* (ASHMEAD). The species known from Honshu are a typical Palearctic lot. A few of the species, such as *Ephedrus plagiator* (NEES), have a wide range both south and north.

Hokkaido Island – Japan. This is the coldest and northern-most island of Japan. The aphidiid fauna contains many species (20) but rather few genera and subgenera (6). There are no endemic genera, but 9 apparently endemic species are found in *Monoctonus (M.)* HALIDAY, *Pauesia* QUILIS, *Aphidius* NEES and *Praon* HALIDAY.

The genus *Pauesia* is quite large in Hokkaido and contains nearly one third of the recorded aphidiid species for the island, while the genus *Aphidius* with its 7 species (several still undescribed) contains another third of the recorded species. With the exception of a few species which occur to the south in Far East Asia, most species in Hokkaido point to a strong cool temperate Palearctic connection with very little direct contact even with northern Honshu.

South Korea. There are now 17 species in 9 genera and subgenera known from South Korea, including the little volcanic island called Chejudo or Saishuto Island. There are no apparent endemic genera or subgenera and only a single *Praon* (possibly new) seems endemic. Nearly one half (8) of the species of South Korea are those that occur westward, but not eastward or southward from Korea in the Palearctic region. The subgenus *Lysephedrus* STARÝ, originally described from Europe, occurs here, but was not found in the other areas visited. Several other interesting distributional patterns reveal themselves. For instance, as noted earlier, the genus *Lysiphlebus* FÖRSTER is not known from any of the Far East Asian islands (being replaced there by *Lysiphlebia* STARÝ and SCHLINGER), but in South Korea we find two species, one of holarctic distribution (*L. salicaphis* FITCH) and the other of Oriental distribution (*L. sp. aff. delhiensis* SHARMA and SUBBA RAO). Two other species showing strictly Oriental influences in South Korea are *Pauesia unilachni* (GAHAN) and *Lysiphlebia rugosa* STARÝ and SCHLINGER, the former known also in Taiwan and Hong Kong, and the latter in Hong Kong. On the other hand, species like *Trioxys (Binodoxys) orientalis* STARÝ and SCHLINGER, *Aphidius "gifuensis"* and *Praon orientale* STARÝ and SCHLINGER show relationship to the Japanese Islands and not to the western Palearctic region. The South Korean fauna appears to be mostly Palearctic with some Oriental influences.

Distribution of Genera and Subgenera. The palearctic and oriental parts of Far East Asia harbor only four endemic genera and subgenera of Aphidiidae. These are: *Archaphidus* STARÝ and SCHLINGER, *Bioxys* STARÝ and SCHLINGER, *Lysiphlebia* STARÝ and SCHLINGER and *Trioxys (Fissicaudus)* STARÝ and SCHLINGER. *Archaphidus* and *T. (Fissicaudus)* are from the Oriental part of the area, *Bioxys* is from the Palearctic part (Manchurian subregion), and *Lysiphlebia* occurs in both of the latter regions. The other genera and subgenera are either Palearctic, Holarctic or cosmopolitan.

SYSTEMATICS

This section describes the taxa of aphidiids known in Far East Asia, and gives their distribution, hosts and habitat conditions.

NEW TAXA

The following list gives the new taxa described in this paper. New genera are: *Archaphidus*, *Bioxys* and *Lysiphlebia*. New subgenus is: *Fissicaudus*. New species are: *Archaphidus greenideae*, *Bioxys japonicus*, *Ephedrus (E.) orientalis*, *Lysiphlebia rugosa*, *Monoctonus similis*, *M. woodwardiae*, *Pauesia tropicalis*, *Praon glabrum*, *P. orientale*, *P. quadratum*, *Trioxys (T.) luteolus*, *T. (Binodoxys) brunnescens*, *T. (B.) carinatus*, and *T. (B.) orientalis*.

LIST OF TAXA

KEY TO THE FAR EAST ASIAN GENERA OF APHIDIIDAE (♀♀)

1   Median vein developed throughout, separating radial cell 1 from
     median cell... 2
     Median vein effaced frontally, radial and median cells confluent;
     venation often reduced behind basal vein.................... 3

2(1) Interradial veins effaced (fig. 169)   Praon HALIDAY
      Interradial veins 1 and 2 developed (fig. 158)   Ephedrus HALIDAY

3(2) Radial and median cells confluent, distinctly completed by
      interradial vein 2 along their external margin; interradial vein
      2 sometimes colorless but distinct (figs. 167) 4
      Radial and median cells confluent, open, not completed by inter-
      radial vein 2 along their external margin (fig. 186) 10

4(3) Pterostigmal cell complete (fig. 161)
              Archaphidus STARÝ and SCHLINGER, n. gen.
      Pterostigmal cell incomplete.............................. 5

5(4) Confluent radial and median cells distinctly separated on lower
      margin by intermedian + median veins (fig. 164) 6
      Confluent radial and median cells on the lower margin open, the
      rest of median vein visible only under interradial vein 2 (fig. 179) 9

6(5) Abdominal segments beginning with the 4th one remarkably
      tubiform and telescopic (fig. 53)   Protaphidius ASHMEAD
      Abdominal segments of normal shape, abdomen lanceolate or
      rounded 7

7(6)   Ovipositor sheaths slightly curved upwards.................. 8
Ovipositor sheaths curved downwards, ploughshare-shaped or
slender, gradually narrowing to the apex (fig. 31)*Monoctonus* HALIDAY

8(7)   Carinae on propodeum forming a large, wide, pentagonal areola,
sometimes barely visible on the longitudinal part (figs. 54)
Pauesia QUILIS
Carinae on propodeum forming a narrow, small, central areola
(fig. 62)   *Aphidius* NEES

9(5)   Propodeum entirely areolated (fig. 119). Tergite 1 coarsely rugose,
with central longitudinal carina (fig. 116)
Lysiphlebia STARÝ and SCHLINGER, n. gen.
Propodeum entirely smooth or with two divergent carinae at the
lower part, always without complete areola (fig. 120). Tergite 1
smooth, with more or less developed central tubercle near the base
only, never rugose or with carina (figs. 117)   *Lysiphlebus* FOERSTER

10(3) Radial vein pointlike. Pterostigma large, strongly sclerotized
(fig. 152). Legs strong   *Paralipsis* FOERSTER
Radial vein distinctly developed, always long, never pointlike.
Legs normal 11

11(10) Ovipositor sheaths curved downwards, terminal abdominal sternite sometimes with 1 or 2 prongs 12
Ovipositor sheaths straight or slightly curved upwards, terminal
abdominal sternite without posterior prongs.................. 15

12(11) Terminal abdominal sternite with 2-1 long or short prongs 13
Terminal abdominal sternite without prongs 14

13(12) Terminal abdominal sternite with single, median, upwardly curved prong (fig. 42)   *Bioxys* STARÝ and SCHLINGER, n. gen.
Terminal abdominal sternite with 2 upwardly curved to nearly
straight prongs (figs. 13 & 16)   *Trioxys* HALIDAY

14(12) Radial vein longer than $^2/_3$ of its possible length so that pterostigmal cell is nearly complete (fig. 183). Ovipositor sheaths slightly
curved downwards, their upper part more strongly sclerotized
(figs. 38 & 40)   *Lipolexis* FOERSTER
Radial vein never longer than $^2/_3$ of its possible length. Pterostigmal cell distinctly incomplete. Ovipositor sheaths slightly curved
downwards, more or less ploughshare-shaped or slender
(see *Monoctonus* HALIDAY)

15(11) Notaulices entirely effaced. Propodeum with a wide, more or less
visible areola (fig. 65)   *Diaeretus* FOERSTER

Notaulices at least at the ascendent part of mesoscutum distinct... 16

16(15) Propodeum entirely areolated, with a small central areola (fig. 62)
         Diaeretiella FOERSTER

Propodeum smooth or with 2 divergent carinae in the lower part.
         (see *Lysiphlebus* FOERSTER)

GENUS APHIDIUS NEES

Incubus SCHRANK, 1802, Fauna boica, 2:315 (Type species: *Ichneumon aphidum* LINNÉ) (Opinions, suppressed)

Aphidius NEES, 1818, Nov. Acta Acad. Caes. Leop. Car. 9:302 (Type species: *Aphidius avenae* HALIDAY (see Opinion 284, Int. Com. Zool. Nomencl.)).

Theracmion HOLMGREN, 1872, Öfvers. Svensk. Vet. Akad. Forh. 29:99 (Type species: *Theracmion arcticus* HOLMGREN).

Literary Data: SMITH, C. F., 1944, Ohio State Univ. Contr. Zoo. Ent. 6:49. STARÝ, 1958, Acta Faun. Ent. Mus. Nat. Pragae 3:56. STARÝ, 1960, Acta Soc. Ent. Cechosl. 57:238.

Description: Head transverse, as wide as or wider than thorax at tegulae. Occiput margined. Antennae with variable number of segments (13 to 23). Notaulices distinct at the ascendent part of mesoscutum. Propodeum areolated, with narrow small central areola. Fore wing: Pterostigma triangular to nearly lanceolate. Metacarp always longer than width of pterostigma. Pterostigmal cell incomplete. Radial and median cell confluent, separated by interradial vein on external side and by fused intermedian and median vein on the lower side. Hind wing with more or less complete basal cell. Abdomen of female lanceolate, rounded at apex in the male. Ovipositor sheaths comparatively short, slightly curved upwards, sparsely haired.

General Distribution: Apparently cosmopolitan, but some regions appear to be without endemic species.

Bionomics: Parasites of aphids. Pupation occurs inside host. Parasitized aphids (mummies) are various shades of brown, and are often shiny.

Key to the species of *Aphidius* (♀♀)

1   Antenna 20 to 21-segmented. Tentorio-ocular line nearly equal to
    $^2/_3$ or intertentorial line (Parasite of *Tuberolachnus salignus*)
         A. salignae WATANABE

Aphidius absinthii MARSHALL

Aphidius absinthii MARSHALL, 1896, in ANDRÉ Spec. Hym. Eur. et d'Alg. 5:605-6 (♂, England, host). MARSHALL, 1899, Trans. Ent. Soc. London 1899:67 (♂, England, host). STARKE, 1956, Nat. Lusatica 3:91 (Germany, host). MACKAUER, 1961, Beitr. Ent. 11:112 (notes on the type). STARÝ, 1961, Ent. Tidskr. 82:218-21 (♀♂, Czechoslovakia, Eur. part of USSR, hosts). MACKAUER, 1962, Beitr. Ent. 12:645.

Aphidius commodus GAHAN, 1926, Proc. U.S. Nat. Mus. 70(8):3-4, New Synonymy.

This is a member of the group of *Aphidius* species that is characterized by having tentorio-ocular line equal to half of intertentorial line. It differs from other species of this group by the width of the temples and genae, number of antennal segments, characters of the female-genitalia and the host-complex. It is typically a parasite of *Macrosiphoniella* species on *Artemisia* and *Achillea* species.

Description – Female: Head transverse, rounded, smooth, shiny, sparsely haired, wider than thorax at tegulae. Temple $^1/_5$ to $^1/_6$ narrower than transverse eye-diameter. Gena as wide as $^1/_4$ of longitudinal eye-diameter. Eyes large, oval, convergent towards clypeus, sparsely and shortly haired. Interocular line about $^1/_2$ longer than transfacial line, a little shorter than facial line. Clypeus with 7 to 13 long hairs. Tentorio-ocular line equal to

$^1/_2$ of intertentorial line. Antenna 16 to 17-segmented (rarely 15 or 18), filiform, slender, reaching to about the center of abdomen. $F_1 = F_2$, about 3.5 to 4 times as long as wide. Socket-ocular line equal to $^1/_2$ of socket-diameter.

Mesoscutum falling arcuately to pronotum, sparsely haired. Notaulices distinct at the ascendent part, deep, slightly crenulate, but effaced on the disc. Propodeum (fig. 76) areolated, with narrow central areola. Discs of areolae smooth, shiny; upper areola with 3 to 5, lower areola with 1 to 3 hairs on either side. Wing (fig. 175): Pterostigma triangular, about 4 times as long as wide. Metacarp $^1/_4$ to $^1/_3$ shorter than pterostigma. Radial abscissa 1 about twice as long as width of pterostigma. Radial abscissa 2 equal to $^1/_2$ of abscissa 1.

Abdomen lanceolate. Tergite 1 (fig. 72) slender, slightly dilating towards apex, more than 3 times as long as wide at spiracles, starting at about half way back with short central longitudinal carina; more or less coarsely rugose-crenulate at the fore part; with slight lateral impressions beyond spiracular tubercles; nearly smooth and slightly convex on the apical part, sparsely haired; less than half as wide at apex than at spiracles. Distance between spiracular tubercles and apex nearly twice as long as width at spiracules. Spiracular tubercles slightly visible, situated before half the length of the tergite. Genitalia figured (fig. 24). Ovipositor sheaths comparatively slender and narrow.

Coloration extremely variable, being yellow to yellowish-orange and/or brown to brownish-black.

Head brownish-black to entirely yellow, with variously distributed dark to light coloration. Antennae brown, lower part of scape, pedicel, and also usually the base of F_1, yellowish; or the quoted segments are entirely brown and only the lower part somewhat lighter; or scape, pedicel, F_1, F_2 and part of F_3 yellow. Thorax brownish-black to nearly entirely yellow, with variously distributed brownish-black and yellow coloration. Wings light yellowish-brown; venation brown; interradial vein, cubital and anal veins on the fore part, and part of cubito-median vein at the lower side of fused radial, and median cells somewhat colorless, but distinct. Legs entirely yellow to entirely dark brown with light trochanters and bases of tibiae. Abdomen entirely yellow with exception of brown tergite 2 and tergite 3 and ovipositor sheaths, to entirely brown with exception of tergite 1, base of tergite 2, suture between tergite 2 and tergite 3, yellow.

Length of body about 1.8 to 2.8 mm.

Male: Antenna 18 to 19-segmented (rarely 17): Tergite 1 relatively more flattened than in female. Coloration less variable than in female. Head blackish-brown; clypeus, lower part of genae and mouthparts yellow or yellowish-brown to almost entirely brown. Antennae brownish-black. Thorax blackish-brown; prothorax brown. Legs dark brown, trochanters and bases of tibiae more or less yellow. Abdomen: Tergite 1 at the base only or entirely yellow, suture between tergite 1 and tergite 2 yellowish. The rest of abdomen brown.

General Distribution: Europe, South Korea, Japan, Hong Kong and Taiwan.

Material Examined (32 ♂♀ specimens). South Korea: Seoul, May 9, 1961, ex. *Macrosiphoniella yomogifoliae* on *Artemisia manischmidtiana*, 19 ♂♀ (E. I. SCHLINGER); Seoul, May 9, 1961, ex. *Macrosiphoniella formosartemisiae* on *Artemisia manischmidtiana*, 1♂ (E. I. SCHLINGER); Seoul, May 9, 1961, ex. *Macrosiphoniella sanborni* on *Chrysanthemum* sp., 1♂ (E. I. SCHLINGER). Japan: Fukuoka, April 17, 1961, ex. *Macrosiphoniella sanborni* on *Chrysanthemum* sp., 8 ♂♀ (E. I. SCHLINGER). Taiwan: Taipei (=Taihoku), Nov. 1921, ex. *Macrosiphoniella tanacetaria*, 1 ♀ (R. TAKAHASHI, USNM) det. by GAHAN as "*Aphidius commodus* GAHAN": Taipei, Oct. 16, 1926, ex. "*Macrosiphum* sp." 1♂ (R. TAKAHASHI). Hong Kong: Kowloon, Feb. 25, 1961, ex. *Macrosiphoniella sanborni* on *Chrysanthemum* sp., 2 ♂♀ (E. I. SCHLINGER).

Type Specimens:

1. *Aphidius absinthii* MARSHALL, holotype ♂, BMNH, Type Hym. 3.c. 188. England.

2. *Aphidius commodus* GAHAN, type ♀, Taihoku, Formosa, ex. *Macrosiphoniella formosartemisiae* TAK., USNM, cat. no. 28985.

Habitat: All specimens collected on *Chrysanthemum* sp. were taken in areas of "gardens" or escaped ornamental plantings, hence other ecological data are unavailable. The specimens collected by SCHLINGER on *Artemisia* in Seoul, South Korea were likewise collected in the Forestry Botanical Garden, and the plants were obviously planted in an artificial environment when sampled.

Hosts: (Unrevised Literature Data).

    Aphidae sp.: STARKE, 1956, on *Artemisia* sp., Germany.

    Macrosiphoniella absinthii: MARSHALL, 1896, 1899, on *Artemisia absinthium*, England.

Macrosiphoniella formosartemisiae TAK.: GAHAN, 1926, Taiwan.

Hosts: (Original and Revised Literature Data).

Macrosiphoniella absinthii (L.): STARÝ, 1961, on *Artemisia absinthium*, Czechoslovakia and USSR (Eur. part).

Macrosiphoniella artemisiae (B.D.F.): STARÝ, 1961, on *Artemisia vulgaris*, Czechoslovakia.

* *Macrosiphoniella formosartemisiae* TAKAH: On *Artemisia manischmidtiana*. South Korea.

Macrosiphoniella millefolii (DEG.): STARÝ, 1961, on *Achillea millefolium*, Czechoslovakia and USSR (Eur. part).

* *Macrosiphoniella sanborni* (GILLETTE): On *Chrysanthemum* sp., Japan, South Korea and Hong Kong.

Macrosiphoniella staegeri HRL.: STARÝ, 1961, on *Centaurea stoebe*, Czechoslovakia.

* *Macrosiphoniella tanacetaria* KOCH: Taiwan.

* *Macrosiphoniella yomogifoliae* TAKAH.: On *Artemisia manischmidtiana*, South Korea.

Macrosiphoniella spp.: STARÝ, 1961, on *Achillea sudetica*, *Anthemis tinctoria*, USSR (Crimea), *Artemisia campestris*, Czechoslovakia.

Host – Specificity: Oligophagous, being apparently restricted to various species of *Macrosiphoniella* that occur on aromatic plants such as *Artemisia* and *Achillea* species. The mummified aphids are shiny brown.

Note: The examination of specimens *A. commodus* GAHAN (as det. by GAHAN) showed that this species is identical with *A. absinthii* MARSH. so that for priority reasons we hereby place *A. commodus* in synonymy with *A. absinthii*.

Aphidius areolatus ASHMEAD

Aphidius areolatus ASHMEAD, 1906, Proc. U.S. Nat. Mus. 30:189-90 (♀♂, Japan). WATANABE, 1957, Ins. Mats. 21:2 (notes on the type).

The remarkably short antennae and number of antennal segments both in the female and male relate this species to the European *A. setiger* MACKAUER. It differs from *A. setiger* by having the tentorio-ocular line equal to $^1/_3$ of intertentorial line, and by having less convergent eyes.

Description – Female: Head (fig. 75) transverse, smooth, shiny, comparatively densely, short haired, nearly evenly narrowed beyond eyes, $^1/_4$ to $^1/_5$ wider than thorax at tegulae. Temple $^1/_3$ narrower than transverse eye-diameter. Gena equal to $^1/_5$ to $^1/_6$ longitudinal eye-diameter. Eyes very large, widely oval, strongly convergent towards clypeus and with sparse short hairs. Interocular line $^1/_2$ to twice as long as transfacial line, a little shorter than facial line. Clypeus with about 10 to 15 long hairs. Tentorio-ocular line about $^1/_3$ of intertentorial line, or a little longer. Antenna 13 to 14-segmented, filiform, densely haired, remarkably short and strong, hardly as long as head and thorax combined. F_1 equal to F_2, twice as long as wide. Socket-ocular line about $^1/_2$ of socket-diameter or somewhat longer.

Mesoscutum slightly elevated above prothorax, without covering it when viewed laterally, smooth with comparatively dense long hairs along margins and on the effaced notaulices on the disc. Notaulices wide and crenulate at the ascendent part, effaced on the disc. Propodeum (fig. 69) distinctly areolated, with narrow, central areola. Upper areola with 10 to 13, lower with 3 to 5 long hairs. Discs of areolae nearly smooth. Wing (fig. 167): Pterostigma more than 4 times as long as wide, longer than metacarp. Radial abscissa 1 equal to abscissa 2.

Abdomen lanceolate. Tergite 1 (fig. 74) 2.5 times as long as wide, slightly dilated backwards, $^1/_5$ wider at apex than at spiracles, with strong, prominent, central longitudinal carina, coarsely rugose on the apical part, with sparse hairs. Spiracular tubercles hardly visible, their neighborhood excavated. Following tergites smooth, shiny, comparatively densely haired. Genitalia as in fig. 26.

Coloration: Head blackish-brown; mouthparts yellowish-brown. Antennae brown. Thorax brownish-black, prothorax lighter. Wings slightly infumated, venation brown. Legs yellowish-brown; coxae, femora and apices of tibiae and tarsi with darkened margin. Tergite 1 yellowish-brown, darkened at apex. Following tergites dark brown, apical sternite yellow. Ovipositor sheaths brown.

Length of body about 2.1 to 2.6 mm.

Male: Antenna 19 to 20-segmented, a little shorter than body. Coloration as in the female except entirely brown apex of abdomen.

General Distribution: Japan.

Material Examined: (65♂♀ specimens). Japan: "Japan", 1♀, type, Cat. No. 7270, USNM (KOEBELE); Fukuoka, April 7, 1961, ex. *Periphyllus testudinacea* on *Acer* sp. 5♂♀ (E. I. SCHLINGER); Kagoshima, April 10, 1961, ex. *Periphyllus* sp. on *Acer palmatum* 4♂♀ (E. I. SCHLINGER); Kyoto, April 22, 1961, ex. *Periphyllus* sp. on *Acer* sp. 5♂♀ (E. I. SCHLINGER); Nopporo, May 26, 1961, ex. *Periphyllus koelreuteriae*, on *Acer* sp., 1♂ (E. I. SCHLINGER); Noboribetsu, May 20, 1961 ex. *Periphyllus* sp. on *Acer* sp., 4 ♂♀ (E. I. SCHLINGER).

Type: ♀, Japan, Cat. No. 7270 (KOEBELE, in USNM).

Habitat: This species was invariably encountered as an arboreal species on *Acer* sp. throughout Japan, both in forested regions as well as in Botanical Gardens. What was presumably this species was collected in large numbers from *Periphyllus* sp. on *Acer* sp. in Seoul, South Korea, but the sample was 100 percent hyperparasitized. Similar habitats in Taiwan failed to show the presence of *A. areolatus*, and the *Periphyllus* sp. were in heavy abundance and causing considerable damage to *Acer* sp. at the time of collection in March 1961.

New Hosts:

* *Periphyllus koelreuteriae* TAKAH: On *Acer* sp., Japan
* *Periphyllus testudinatus* (FERNIE): On *Acer* spp., Japan
* *Periphyllus* sp. on *Acer palmatum* and *Acer* sp., Japan

Note: Although *P. koelreuteriae* is recorded here as a host of this parasite on *Acer* sp., it should be noted that SCHLINGER observed this same host aphid on *Koelreuteria* in Asakawa, Japan in such heavy numbers as to cause considerable damage. Therefore it may be that this parasite is selecting *Periphyllus* sp. on *Acer*.

Host-specificity: Apparently specific to species of *Periphyllus*.

Note: The parasitized aphids (mummies) are yellowish-brown to golden brown in color.

Aphidius "gifuensis ASHMEAD", group

Aphidius gifuensis ASHMEAD, 1906, Proc. U.S. Nat. Mus. 30:188

(♀, Japan): WATANABE, 1957, Ins. Mats. 21:2 (notes on type).

This species group differs from its relatives by 17 to 18-segmented antennae in the female and in the tentorio-ocular line, equal to $1/3$ of intertentorial line, and in coloration. Judging from the criteria used, our material seems to belong to only one species, but it is unusual for any *Aphidius* species to parasitize such different groups of aphids as *Myzus* and *Acyrthosiphon*. For this reason we have classified *A. gifuensis* as a species group and hope the problem of specificity can be solved later as more material becomes available.

Description – Female: Head transverse, rounded, smooth, shiny, sparsely haired, wider than thorax. Temple about $1/3$ to $1/5$ narrower than transverse eye-diameter. Gena as wide as $^1/_5$ of longitudinal eye-diameter. Eyes large, widely oval, sparsely haired, strongly convergent towards clypeus. Interocular line, slightly longer than transfacial line, or nearly so, a little shorter than facial line. Clypeus with about 8 to 10 long hairs. Tentorio-ocular line equal to $^1/_3$ of intertentorial line, or a little longer. Antenna 17 to 18-segmented, a little shorter than the body. $F_1 = F_2$, 4 to 5 times as long as wide. Socket-ocular line, equal to half of socket-diameter.

Mesoscutum smooth, shiny, with sparse long hairs along margins and effaced notaulices on the disc. Notaulices wide and crenulate at the ascendent part, effaced on the disc; fore margin slightly prominent. Propodeum (fig. 71) areolated, with narrow central areola, carinae strongly prominent. Upper areola with 5 to 8, lower with 2 to 4 hairs. Wing (fig. 174): Pterostigma 4 times as long as wide, metacarp equal or longer than pterostigma. Radial abscissa 1 equal to abscissa 2 or longer.

Abdomen lanceolate. Tergite 1 (fig. 79) 3.5 to 4 times as long as wide, slender, slightly dilating towards apex, with central longitudinal carina, rugose, sparsely haired. Spiracular tubercles poorly visible. Following tergites smooth, shiny, sparsely haired. Genitalia as figured (fig. 22).

Coloration variable. Head blackish-brown; lower part of genae, clypeus and mouthparts yellow. Antennae dark brown; scape, lower part of pedicel and basal part of F_1 yellow, sometimes F_1 entirely, and base of F_2 yellow. Thorax with variously distributed brown and yellowish-brown coloration; usually mesoscutum, metanotum and part of propodeum brown, the rest yellowish-brown; or nearly all the thorax yellowish-brown; or nearly brown, with yellowish-brown coloration on prothorax and propleurae. Wings hyaline, venation brown. Legs yellow, apices of tarsi darkened. Tergite 1 yellow, tergite 2 yellow at base and with 2 brown lateral spots, suture between tergites 2 and 3 with wide yellow band. The following tergites brown; area near apex of abdomen with large yellow lateral spots. Ovipositor sheaths brown.

Length of body about 2.1 to 2.4 mm.

Male: Antenna 19 to 20-segmented. Head nearly entirely brown. Antennae brown. Legs yellowish-brown, with darkened tinge. Tergite 1 and base of tergite 2 yellow, suture between tergites 2 and 3 with yellow band, the rest dark brown.

General Distribution: South Korea, Japan and Taiwan.

Material Examined: (86♂♀ specimens). South Korea: Chejú-do Island, May

5, 1961, ex. *Myzus persicae* on *Citrus sinensis*, 1 ♂ (E. I. SCHLINGER); Chejú-do
Island, May 6, 1961, ex. *Myzus persicae* on *Malva* sp., 25♂♀ (E. I. SCHLINGER).
Japan: Maruyama, Sapporo, May 23, 1961, ex. *Acyrthosiphon* sp. on *Corydalis
platycarpa*, 11♂♀ (E. I. SCHLINGER); Kagoshima, April 11, 1961, ex. *Acyrthosiphon* sp., on *Delphinium* sp., 3 ♂♀ (E. I. SCHLINGER); Fukuoka, April 6, 1961,
ex. *Macrosiphum* sp. on *Spiraea thunbergi*, 4 ♂♀ (E. I. SCHLINGER); Fukuoka,
April 8, 1961, ex. *Macrosiphum* sp. on *Malva* sp., 6 ♂♀ (E. I. SCHLINGER).
Taiwan: Taipei, March 6, 1961, ex. *Macrosiphum rosaeibarae* on *Rosa* sp.,
2 ♂♀ (E. I. SCHLINGER); Taichung, March 12, 1961, ex. *Myzus persicae* on
Nicotiana sp., 27 ♂♀ (E. I. SCHLINGER); Yung Jean, March 11, 1961, ex. *Myzus
persicae* on *Verbena phlogiflora*, 5 ♂♀ (E. I. SCHLINGER); Wulai, March 21, 1961,
ex. *Myzus* sp., 1 ♀ (E. I. SCHLINGER); Wulai, March 21, 1961, ex. *Capitophorus*
sp., 1 ♀ (E. I. SCHLINGER).

Type: Lectotype ♀, Gifu, Japan (Y. NAWA), Cat. No. 7267, USNM (*A.
gifuensis* ASHMEAD).

Habitat: This species seemed to be omnipresent in Japan, Taiwan and on
Chejú-do Island in South Korea, but was never exceedingly abundant. It
was most often encountered as a parasite of *Myzus persicae* on several
herbaceous plants. It was found in forested areas as well as open to woody
steppe regions.

Hosts: (Unrevised Literature Data)

　Aphidae sp.: ASHMEAD, 1906, Japan, WATANABE, 1957, on *Euphrasia
　inumai* TAKEDA.

Hosts: (Original and Revised Literature Data)

* *Acyrthosiphon* sp.: on *Corydalis platycarpa*, Japan.
* *Acyrthosiphon* sp.: On *Delphinium* sp., Japan.
* *Capitophorus* sp.: On unknown plant, Taiwan.
* *Macrosiphum* sp.: On *Malva* sp., Japan.
* *Macrosiphum* sp.: On *Spiraea thunbergi*, Japan.
* *Macrosiphum rosaeibarae* MATS.: On *Rosa* sp., Taiwan.
* *Myzus persicae* SULZ.: On *Citrus sinensis*, S. Korea. On *Malva* sp.,
　S. Korea. On *Nicotiana* sp., Taiwan. On *Verbena phlogiflora*, Taiwan.
* *Myzus* sp.: On unknown plant, Taiwan.

Note: The mummified aphids are yellowish-brown.

Aphidius salicis HALIDAY

Aphidius (Aphidius) salicis HALIDAY, 1834, Ent. Mag. 2:102 (♀♂, England, host). THOMSON, C. G., 1895, Opusc. entomol. 20:2337 (♀, Sweden). MARSHALL, 1896, in ANDRÉ, Spec. Hym. Eur. et d'Alg. 5:594-5 (♀♂, England, host). MARSHALL, 1899, Trans. ent. Soc. London 1899: 60 (♀♂, England, host). HAVILAND, 1921, Quart. J. microsc. Sci. 65: in 101-127 (England, host)[?]. HAVILAND, 1922, Quart. J. microsc. Sci. 66: in 321-8 (England, host) [?]. SCHIMITSCHEK, 1936, Z. ang. Ent. 22:564 (Austria, host) [?]. MACKAUER, 1961, Beitr. Ent. 11: 131-2 (notes on type, synonymy). MACKAUER, 1962, Beitr. Ent. 12: 648.

Aphidius dauci MARSHALL, 1896, in ANDRÉ, Spec. Hym. Eur. et d'Alg. 5:601-2 (♀♂, England, host). MARSHALL, 1899, Trans. ent. Soc. London 1899:64-5 (♀♂, England, host). QUILIS, M. P., 1934, Eos 10:12 (Czechoslovakia, host) [?]. LUZHETZKI, 1960, Par. tlej Uzbekistana, p. 132-3 (♀♂, USSR – Uzbekistan) [?]. MACKAUER, 1961, Beitr. Ent. 11:118 (notes on type).

This species is similar to *A. areolatus* ASHM., as regards the number of antennal segments in female, but differs from that species by having smaller eyes, the relation of intertentorial and tentorio-ocular lines, and the host-complex.

Description – Female: Head transverse, rounded, smooth, shiny, wider than thorax at tegulae. Temple a little narrower than transverse eye-diameter. Gena as wide as $1/5$ of longitudinal eye-diameter. Eyes small, oval, slightly prominent, sparse and short haired, slightly convergent towards clypeus. Interocular line, $1/2$ longer than transfacial line, a little shorter than facial line. Clypeus with about 10 long hairs. Tentorio-ocular line equal to $1/3$ of intertentorial line. Antenna 13-segmented (more rarely 14), filiform, not quite as long as head and thorax together. F_1 equal to F_2, nearly 3 times as long as wide. Socket-ocular line $1/3$ shorter than socket-diameter.

Mesoscutum gibbous, vertically falling to prothorax, smooth, shiny, with sparse long hairs along margins and effaced notaulices on the disc. Notaulices narrow, feebly crenulate, distinct at the ascendent part and effaced on the disc. Propodeum (fig. 73) areolated; upper areola with 7, lower with 2 hairs. Wing (fig. 173): Pterostigma more than 4 times as long as wide. Metacarp equal to $1/2$ of pterostigma. Radial abscissa 1 twice as long as abscissa 2.

Abdomen lanceolate. Tergite 1 (fig. 78) more than twice as long as wide at spiracles, nearly parallel-sided, with short central longitudinal carina along which is excavated laterally, rugose, sparsely haired. Following tergites smooth, shiny, sparsely haired. Genitalia as figured (fig. 25).

Coloration: Head black, mouthparts brown. Antennae brownish-black. Thorax black. Wings almost hyaline, venation brown. Legs dark brown, trochanters and base of tibiae yellow. Tergite 1 brown. Abdomen brown, suture between tergites 2 and 3 yellow.

Length of body about 1.4 to 1.7 mm.

Male: Antenna 16 to 17-segmented, otherwise as described for the female.

General Distribution: Europe, South Korea, Japan, and Taiwan.

Material Examined: (68♂♀ specimens). Japan. Noboribetsu, May 19, 1961, ex. *Cavariella* sp. on *Salix* sp. 37 ♂♀ (E. I. SCHLINGER). South Korea: Seoul, May 9, 1961, ex. *Cavariella salicicola* on *Salix* sp., 8 ♂♀ (E. I. SCHLINGER). Taiwan: Taipeito, March 19, 1961, ex. *Cavariella araliae* on *Aralia* sp., 23 ♂♀ (E. I. SCHLINGER).

Type Specimens:

1. *Aphidius salicis* HALIDAY, Lectotype ♀, BMNH, Type Hym. 3.c. 102. England. Allotype, BMNH, Type Hym. 3.c. 102 (on same pin as lectotype).

2. *Aphidius dauci* MARSHALL, Holotype. ♀, BMNH, Type Hym. 3.c. 90. South Devonshire.

Habitat: This species was encountered in Far East Asia only on *Cavariella* species, either on *Salix* sp. or *Aralia* sp., but always in sandy soil near stream beds. The fact that many specimens of *Cavariella araliae* were encountered in a similar habitat in Hong Kong, parasitized only by *Ephedrus plagiator*, might indicate that *A. salicis* does not occur in South China.

Hosts: (Unrevised Literature Data)

Aphidae sp.: HALIDAY, 1834, on *Salix* sp., England.

Aphidae sp.: MARSHALL, 1896, on *Daucus carota* and *Crithmum maritimum*.

Aphis farinosa GMEL.: HAVILAND, 1922, 1931, England. SCHIMITSCHEK, 1936, Austria.

Cavariella pastinacae L.: MARSHALL, 1896, on *Pastinaca sativa* and *Apium graveolens*.

(?) *Hayhurstia atriplicis* (L.): QUILIS M. P., 1934, on *Chenopodium vulvaria*, Czechoslovakia.

Hosts: (Original and Revised Literature Data)

Aphis (Toxopterina) lambersi BOERNER: On *Daucus carota*, Czechoslovakia.

★ *Cavariella araliae* TAKAH: On *Aralia* sp., Taiwan.

★ *Cavariella salicicola* (MATS.): On *Salix* sp., South Korea.

Cavariella spp.: On *Daucus carota*, Czechoslovakia. On *Anthriscus sil-*

vestris, Czechoslovakia. On *Selinum carvifolia*, Czechoslovakia. On *Angelica silvestris*. Czechoslovakia.

★ *Cavariella* sp.: On *Salix* sp., Japan.

Host-specificity: Primarily an oligophagous parasite of *Cavariella* species, but several other species are questionably recorded as hosts.

Aphidius salignae WATANABE

Aphidius salignae WATANABE, 1941, Ins. Mats. 15:81-3 (♀♂, Japan, host). YASUMATSU, ISHIHARA, & MORITSU, 1946. Mushi, 17: in 9-12 (Japan, host).

This species is related to *A. cingulatus* RUTHE, from which it differs by characters on the head, length of notaulices, shape of ovipositor sheaths, coloration, and the host-complex.

Description – Female: Head transverse, smooth, shiny, comparatively densely haired, strongly narrowed beyond eyes, wider than thorax. Temple about $1/_6$ narrower than transverse eye-diameter. Gena as wide as $1/_3$ of longitudinal eye-diameter. Eyes of middle size, nearly hemispherical, strongly prominent, sparsely haired, slightly convergent towards clypeus. Interocular line a little longer than transfacial line, a little shorter than facial line. Clypeus with about 15 long hairs. Tentorio-ocular line nearly $1/_3$ of the length of intertentorial line. Antenna 20 to 21-segmented, filiform, reaching nearly half the length of abdomen. F_1 equal to F_2, twice as long as wide. Socket-ocular line equal to socket-diameter.

Mesoscutum strongly prominent, nearly covering prothorax as seen from side, smooth, shiny, comparatively dense and long haired over nearly all surface and on central lobe. Notaulices coarsely crenulate, wide, distinct at ascendent part and nearly to $1/_2$ of mesoscutum and then as feeble impressions nearly up to praescutellar groove; fore margin strongly prominent, so that central lobe has an angular shape as seen from above. Propodeum (fig. 70): with narrow central areola and irregular carinae on lower part; discs of areolae densely haired, with about 20 hairs. Wing (fig. 171): Pterostigma nearly 3 times as long as wide. Metacarp somewhat longer than pterostigma. Radial abscissa 1 about $1/_2$ the length of pterostigma, and is equal to abscissa 2.

Abdomen lanceolate. Tergite 1 (fig. 77) 3 times as long as wide at spiracles, slightly dilating backwards, with feeble lateral impressions beyond spiracular tubercles, with central longitudinal carina, with a central longitudinal impression on the hind part along which there are tuberculiform protuberances; surface shiny, rugose, comparatively densely haired. Following tergites smooth, shiny, comparatively dense and long haired. Genitalia as figured (fig. 23). Ovipositor sheaths strongly narrowed towards apex.

Coloration: Head yellow with variously distributed brown coloration, usually the upper part of frons, vertex, occiput and part of temples are brown. Antennae brown, scape at the lower part yellow, the base of F_1 and F_2 sometimes lighter colored. Prothorax yellow. Mesoscutum yellow with brown spots on the lobes. The rest of thorax yellow to brown. Wings hyaline, venation brown. Tegulae yellow. Legs yellow, tarsi obscured. Tergite 1 brown, remaining tergites brown, suture between tergite 2 and 3 broadly yellow.

Length of body about 2.4 to 2.8 mm.

Male: Antenna 23 to 25-segmented. Coloration as in the female, but the brown coloration distributed more widely.

General Distribution: Japan, Taiwan, and California, USA.

Material Examined: (36 ♂♀ specimens). Japan: Ibusuki, April 13, 1961, ex. *Tuberolachnus salignus* on *Salix* sp., 31♂♀ (E. I. SCHLINGER), Gifu, October 1902, 2 ♂ (Nawa), red label-paratype No. 7268 USNM, 200. These two males are labeled "Yanagi" (=willow) in a series of paratype specimens of *A. japonicus* ASHM. WATANABE (1957, p. 2) correctly distinguished these specimens as *salignae*. Taiwan: Taipei, March 24, 1961, ex. *Tuberolachnus salignus* on *Salix* sp., 3♂♀ (E. I. SCHLINGER). California: Many specimens have been reared (1962-64) ex. *Tuberolachnus salignus* on *Salix* sp. from Riverside, Del Mar and San Luis Obispo coastal areas (E. I. SCHLINGER, R. VAN DEN BOSCH).

Holotype ♀: Sapporo, Japan, July 23, 1938, ex. *Tuberolachnus salignus* on Salicaceae, C. WATANABE (Hokkaido Univ. Collection).

Allotype: Same data as holotype.

Habitat: This species has invariably been collected in *Salix*-inhabited areas, and usually in association with tree-like *Salix* sp. in Taiwan, Japan and California. The record from California is the first nearctic record for this species. It is important to note that in the drier areas of California, where the host aphid has often been a minor pest on *Salix* sp., we have only rarely located any parasites. It was in the cool and more moist coastal areas where we found this parasite in California, and we presume this represented an undetected natural occurrence rather than an accidental introduction. Presumably this species occurs or has occurred north along the coast into Canada and even Alaska, thence west along the coast of USSR and south to Japan.

Hosts: (Unrevised Literature Data).

Tuberolachnus salignus GMEL: WATANABE, 1941, on Salicaceae, Japan; YASUMATSU, et. al. 1946, Japan.

Hosts: (Original Data)

* *Tuberolachnus salignus* GMEL: On *Salix* sp., Japan, Taiwan and California.

Host-specificity: Apparently this parasite is specific on its rather unique host, *Tuberolachnus salignus* GMELIN.

Note: The mummified aphids are rather dull black, sometimes with a dark brown tinge.

Undetermined species of *Aphidius* NEES

We have been able to distinguish at least seven more species of Far East Asian *Aphidius* which are at present definable but nameless and which might represent new species. More specimens are needed to clarify their status, but partial descriptions are given here for future workers.

Aphidius species – No. 1

Description – Female: Head transverse, rounded, sparsely haired, wider than thorax at tegulae. Temple $^1/_4$ narrower than transverse eye diameter. Gena as long as $^1/_5$ of longitudinal eye-diameter. Eyes of medium size, oval, sparsely haired. Interocular line $^1/_2$ longer than transfacial line, shorter than facial line. Clypeus with 8 long hairs. Tentorio-ocular line equal to $^1/_3$ of intertentorial line. Antenna 16-segmented, filiform, reaching to about half the length of abdomen. F_1 equal to F_2, 3 times as long as wide.

Mesoscutum smooth, shiny, with sparse long hairs along margins and on effaced notaulices on the disc. Notaulices distinct on the ascendent part and effaced on the disc. Propodeum with narrow central areola, upper areola with 3, lower with 2 hairs. Wing: Pterostigma triangular, 3 times as long as wide, longer than metacarp. Radial abscissa 1 longer than abscissa 2.

Abdomen lanceolate. Tergite 1 slender, 3 times as long as wide, slightly dilating backwards, with central longitudinal carina, rugose at the fore part and nearly smooth at the apex, convex, with sparse hairs.

Coloration: Head brown, face, lower part of genae, clypeus and mouthparts yellow. Antennae brown, apex of pedicel and narrow base of F_1 yellowish. Thorax brown, prothorax light brown. Wings hyaline, venation brown.

Legs brown, trochanters yellow. Abdomen brown, basal part of tergite 1 and base of tergite 2 yellow.

Length of body about 2 mm.

Male: Unknown

Material Examined: (1 ♀ specimen). Taiwan: Taipei, March 5, 1961, ex. *Macrosiphum rosaeibarae* on *Rosa* sp., 1 ♀ (E. I SCHLINGER).

Note: The mummified aphids are yellowish-brown.

Aphidius species – No. 2

Description – Female: Head transverse, smooth, shiny, rounded, sparsely haired, wider than thorax at tegulae. Temple nearly equal to transverse eye-diameter. Gena as long as $1/_6$ of longitudinal eye-diameter. Eyes of middle size, oval, sparsely haired, convergent towards clypeus. Interocular line nearly equal to twice that of transfacial line, a little shorter than facial line. Clypeus with about 10 hairs. Tentorio-ocular line equal to $1/_3$ of intertentorial line. Antennae broken. F_1 equal to F_2, 3 times as long as wide. Socket-ocular line equal to $1/_3$ of socket-diameter.

Mesoscutum with sparse long hairs along margins and effaced notaulices. Notaulices distinct at the ascendent part of mesoscutum, effaced on the disc. Propodeum with narrow central areola, upper areola with 6, lower with 3 hairs. Wing: Pterostigma more than 4 times as long as wide, longer than metacarp. Radial abscissa 1 longer than abscissa 2.

Abdomen broken.

Coloration: Head brownish-black, lower part of genae, clypeus, the neighborhood of antennal sockets, yellow. Scape, pedicel and basal $1/_2$ of F_1 yellow. Base of mesoscutum, prothorax, mesopleurae and metapleurae yellowish brown, the rest of thorax dark brown. Wings hyaline, venation brown. Legs yellow.

Length of body about 2.1 mm.

Male: Similar to female except with darker coloration and tergite 1, base of tergite 2 and suture between tergites 2 and 3 yellow, the rest of abdomen brown.

Material Examined: (2 ♂♀ specimens). Japan: Sapporo, May 24, 1961, ex. *Acyrthosiphon* sp. on *Quercus* sp., 1♀, 1♂, (E. I. SCHLINGER).

Note: The mummified aphids are yellowish-brown.

Aphidius species – No. 3

Description – Female: Head transverse, rounded, smooth, shiny, sparsely haired. Temple nearly equal to transverse eye-diameter. Gena as wide as $^1/_5$ of longitudinal eye-diameter. Eyes of medium size, sparsely haired, convergent towards clypeus. Interocular line equal to nearly twice that of transfacial line, a little shorter than facial line. Clypeus with about 10 hairs. Tentorio-ocular line equal to $^1/_3$ of intertentorial line. Antennae broken. F_1 equal to F_2, more than 3 times as long as wide. Socket-ocular line shorter than half of socket-diameter.

Mesoscutum with sparse long hairs along margins and effaced notaulices on the disc. Notaulices distinct at the ascendent part, effaced on the disc. Propodeum with narrow central areola, upper areola with 8, lower with 3 hairs. Wing: Pterostigma 5 times as long as wide, longer than metacarp. Radial abscissa 1 longer than abscissa 2.

Abdomen lanceolate. Tergite 1 slender, slightly dilating backwards, with feeble central longitudinal carina, granulo-rugose, sparsely haired. Spiracular tubercles hardly visible.

Coloration: Head brownish-black; lower part of genae, clypeus and mouthparts yellow. Antennae with basal half yellowish-brown, remainder brown. Thorax yellowish-brown; mesoscutum and metanotum dark brown. Wings hyaline, venation brown. Legs yellowish-brown. Tergite 1 and base of tergite 2 yellow, suture between tergites 2 and 3 yellow, remainder brown.

Length of body about 2.4 mm.

Male: Unknown

Material Examined: (1♀ specimen). South Korea: Seoul, May 15, 1961, ex. *Macrosiphum rosae* on *Rosa* sp., 1♀, (E. I. SCHLINGER).

Note: The mummified aphids are brown.

Aphidius species – No. 4

Description – Female: Head transverse, smooth, shiny, sparsely haired, wider than thorax. Temple a little narrower than transverse eye-diameter. Gena equal to $^1/_5$ of longitudinal eye-diameter. Eyes of medium size, sparsely haired, convergent towards clypeus. Interocular line $^1/_2$ longer than transfacial line, shorter than facial line. Clypeus with 9 long hairs. Tentorio-ocular line equal to $^1/_3$ of intertentorial line. Antenna 17-segmented, filiform. F_1 equal to F_2, 4 times as long as wide. Socket-ocular line equal to $^1/_2$ of socket diameter.

Mesoscutum smooth, shiny, with long hairs along margins and effaced notaulices on the disc. Propodeum with narrow central areola: upper areola with 7, lower with 3 hairs. Wing: Pterostigma 4 times as long as wide, equal to metacarp. Radial abscissa 1 a little shorter than abscissa 2.

Mesoscutum smooth, shiny, with long hairs along margins and effaced notaulices on the disc. Notaulices rather wide, crenulate at the ascendent part and effaced on the disc. Propodeum with narrow central areola: upper areola with 7, lower with 3 hairs. Wing: Pterostigma 4 times as long as wide, equal to metacarp. Radial abscissa 1 a little shorter than abscissa 2.

Abdomen lanceolate. Tergite 1 more than 3 times as long as wide, slender, dilating towards apex, with feeble lateral impressions, prolongately granulate-rugose, sparsely haired. Spiracular tubercles hardly distinguishable.

Coloration: Head blackish-brown; face, clypeus, lower part of genae yellow. Basal part of antennae yellow, the rest brown. Thorax yellowish-brown, lower portion yellow. Wings hyaline, venation light brown. Tergite 1 yellow, following tergites yellow with brown lateral spots; suture between tergites 2 and 3 white. Ovipositor sheaths yellow.

Length of body about 2.4 mm.

Male: Unknown

Material Examined: (1♀ specimen). Japan: Kyoto, April 12, 1961, ex. *Rhopalosiphoninus deutzifoliae* on *Spiraea* sp., 1♀ (E. I. SCHLINGER).

Note: The mummified aphids are brown.

Aphidius species – No. 5

Description – Male: Head rounded, smooth, shiny, sparsely haired, wider than thorax at tegulae. Temple nearly $^1/_3$ narrower than transverse eye-diameter. Gena equal to $^1/_4$ of longitudinal eye-diameter. Eyes small, sparsely haired, convergent towards clypeus. Interocular line a little shorter than twice that of transfacial line, just a little shorter than facial line. Clypeus with about 11 hairs. Antenna 19-segmented. Tentorio-ocular line as $^1/_3$ of intertentorial line.

Mesoscutum with sparse hairs along margins and effaced notaulices on the disc. Notaulices deep, rugose at the ascendent part and effaced on the disc. Propodeum areolated: upper areola with 4, lower with 2 hairs. Wing:

Pterostigma more than 3.5 times as long as wide, equal to metacarp. Radial abscissa 1 longer than abscissa 2.

Abdomen rounded at apex. Tergite 1 nearly parallel-sided, more than 3 times as long as wide, rugose-granulate, with short central longitudinal carina and feeble lateral impressions.

Coloration: Head blackish-brown, lower part of clypeus and mouthparts yellow. Antennae brown. Thorax blackish-brown, prothorax brownish-yellow or only mesoscutum and metanotum dark brown and the rest of thorax yellowish-brown. Wing venation brown. Legs yellowish-brown, trochanters and bases of tibiae yellow. Tergite 1 brownish-yellow to yellow, suture between tergites 2 and 3 yellow, the rest of abdomen brown.

Length of body about 2.2 mm.

Female: Unknown

Material Examined: (2 ♂♂ specimens). Japan: Sapporo, May 27, 1961, ex. *Acyrthosiphon* sp. on *Syringa reticulata*, 2 ♂♂ (E. I. SCHLINGER).

Note: The mummified aphids are brown.

Aphidius species – No. 6

Description – Female: Head transverse, rounded, sparsely haired, wider than thorax at tegulae. Temple about $1/3$ narrower than transverse eye-diameter. Gena equal to $1/3$ of longitudinal eye-diameter. Eyes large, widely oval, convergent towards clypeus. Interocular line $1/2$ longer than trans-facial line, a little shorter than facial line. Clypeus with about 13 long hairs. Tentorio-ocular line equal to $1/3$ of intertentorial line. Antenna 17-segmented, filiform, F_1 equal to F_2, 5 times as long as wide. Socket-ocular line equal to $1/3$ of socket-diameter.

Mesoscutum with sparse long hairs along margins and effaced notaulices on the disc. Notaulices deep, crenulate at the ascendent part and effaced on the disc. Propodeum with narrow central areola, upper areola with 10, lower with 4 long hairs. Wing: Pterostigma more than 4 times as long as wide, longer than metacarp. Radial abscissa 1 a little longer than abscissa 2.

Abdomen lanceolate. Tergite 1 slender, 3.5 times as long as wide, rugose-granulate, with short central longitudinal carina and feeble lateral impressions, with longitudinal central impression on the hind part and lateral protuberances along it, sparsely haired.

Coloration: Head dark brown, the neighborhood of mandibles and mouthparts yellow. Scape, pedicel and base of F_1 yellow, the rest brown. Mesoscutum and metanotum dark brown, the rest of thorax yellowish-brown. Venation brown. Legs yellow, praetarsi obscured. Tergite 1, a spot at the base of tergite 2 and suture between tergites 2 and 3 yellow, the rest brown.

Length of body about 2.4 mm.

Male: Unknown

Material Examined: (1♀ specimen). Japan: Sapporo, May 27, 1961, ex. *Amphorophora* sp. on *Vaccinium*, 1♀ (E. I. SCHLINGER).

Note: The mummified aphid is yellowish-brown.

Aphidius species – No. 7

Description – Female: Head transverse, smooth, shiny, sparsely haired, wider than thorax at tegulae. Temple about $1/3$ narrower than transverse eye-diameter. Gena equal to $1/3$ of longitudinal eye-diameter. Eyes of medium size, widely oval, sparsely haired, convergent towards clypeus. Interocular line about $1/2$ longer than transfacial line, shorter than facial line. Clypeus with 12 long hairs. Tentorio-ocular line equal to $1/3$ of intertentorial line. Antenna 19-segmented, filiform. F_1 equal to F_2, more than 3.5 times as long as wide. Socket-ocular line shorter than half of socket-diameter.

Mesoscutum smooth, shiny, with long sparse hairs along margins and effaced notaulices on the disc. Notaulices narrow, crenulate at the ascendent part, effaced on the disc. Propodeum areolated, upper areola with 8, lower with 5 hairs. Wing: Pterostigma triangular, more than 4 times as long as wide, equal to metacarp. Radial abscissa 1 equal to abscissa 2.

Abdomen lanceolate. Tergite 1 more than 3.5 times as long as wide, slender, slightly dilating towards apex, nearly parallel-sided, coarsely rugose-granulate, with feeble lateral impressions, sparsely haired.

Coloration: Head blackish-brown; mouthparts yellow. Lower part of scape, pedicel, F_1 and base of F_2 yellow. Thorax blackish-brown. Wings almost hyaline, venation brown. Legs yellow, apices of tarsi darkened. Tergite 1 yellow, base of tergite 2 and suture between tergites 2 and 3 yellow. The rest brown.

Length of body about 2.3 mm.

Male: Unknown

Material Examined: (1♀ specimen). Japan: Sapporo, Maruyama, May 29, 1961, ex. *Acyrthosiphon* sp. on *Corydalis platycarpa*, 1♀ (E. I. SCHLINGER).

Note: The mummified aphid is brown.

Aphidius species

Several swept specimens could not be determined because of the contemporary state of our knowledge. These are simply noted here for generic, distributional and seasonal data.

Material Examined: USSR – Primorye: Akademgorodok, env. of Vladivostok, July 16 and 27, 1961 and August 27, 1961, swept from grasses, coll. (TRIAPICYN). Buchta Taungou, Sudzhuchinsky zapovednik, sandy habitats on the seashore (TRIAPICYN). Artjom, env. of Vladivostok, July 15, 1961, (TRIAPICYN). Gorno-Tajozhnaja, 10. VIII. 1961, edge of forest, (TRIAPICYN). Stanciya Sedanka, Vladivostok district, July 27, 1961, bushes, (TRIAPICYN). Stancija Ugolnaya, env. of Vladivostok, July 25, 1961, (CHUVACHINA). Okeanskaya, env. of Vladivostok, Botanical garden, July 11, 1961, (TRIAPICYN). Suputinsky zapovednik, Aug. 1, 1961, (TRIAPICYN).

GENUS ARCHAPHIDUS STARÝ AND SCHLINGER, NEW GENUS

This new genus is rather peculiar and differs from all other known genera of the Aphidiidae by numerous features. The venation of the wing is similar to *Aclitus* FOERSTER, but there are so many other different characters from *Aclitus* that we feel the venation represents clearly a case of convergence rather than a true relationship.

Description: Head transverse. Eyes large. Antennae filiform, consisting of 17 segments. Notaulices effaced. Propodeum partly carinated. Fore wing: Pterostigma triangular. Pterostigmal cell complete. Radial cell 1, 2, and median cell confluent, completed by fused intermedian and median veins and by interradial vein 2. Radial cell 3 nearly complete, median vein nearly reaches wing apex. Cubital cell 2 complete. Hind wing with complete basal cell. Abdomen of female lanceolate. Tergite 1 slender. Ovipositor sheaths curved upwards (?).

Type Species: Archaphidus greenideae STARÝ and SCHLINGER, new species, by present designation.

Archaphidus greenideae STARÝ and SCHLINGER, new species

Description – Female: Head transverse, smooth, shiny, sparsely haired, wider than thorax at tegulae. Occiput margined. Temple about $^1/_6$ narrower than transverse eye-diameter. Gena as wide as $^1/_4$ of longitudinal eye-diameter. Eyes large, oval, prominent frontolaterally, sparsely and shortly haired, slightly convergent towards clypeus. Interocular line a little longer than transfacial line (17:14). Facial line longer than interocular line (22:17). Clypeus slightly prominent, narrow, with 10 long hairs. Tentorio–ocular line about half of intertentorial line. Antenna 17-segmented, filiform, reaching to half of abdomen. Flagellar segments 1 and 2 of equal length, more than 5 times as long as wide. Antennal socket diameter twice as long as socket–ocular line.

Mesoscutum falling rather vertically to the prothorax, without covering it when viewed laterally, smooth, with sparse long hairs along effaced notaulices. Notaulices effaced. Propodeum (fig. 115) with two distinct divergent carinae at the lower part, nearly smooth, sparsely haired. Fore wing (fig. 161): Pterostigma more than 3 times as long as wide. Pterostigmal cell complete. Radial cell 1, 2, and median cell confluent, separated by interradial vein 2 from nearly complete radial cell 3. Hind wing with complete basal cell.

Abdomen lanceolate. Tergite 1 (fig. 118) slender, 3.5 times longer than wide at spiracles, a little wider at apex than at spiracles, prolongately rugose, with feeble central longitudinal carina and sparse long hairs. Spiracular tubercles prominent laterally, situated before middle of tergite. Following tergites smooth, with sparse long hairs. Ovipositor sheaths slightly curved upwards (?) as far as visible.

Coloration: Head yellow, antennae brown except yellow base. Thorax yellow. Wings hyaline, venation light yellowish-brown. Legs light yellow, praetarsi brown. Tergite 1 yellowish-brown. Tergites 2 and 3 yellow, brown laterally, the following tergites banded with brown spots at basal part and on sides.

Length of body about 3.1 mm.

Male: Unknown

Holotype ♀: Taiwan, Sun Moon Lake, March 14, 1961, bred from *Greenidea ficicola* TAKAHASHI on *Ficus* sp., (E. I. SCHLINGER).

Habitat: The single specimen was collected together with several other parasitized aphids which failed to give rise to healthy parasites. The parasitized aphids were found on the underside of *Ficus* leaves in a semi-forested area near a tea plantation. The aphid mummies were all yellowish-brown in color. Pupation occurred inside parasitized aphid.

GENUS BIOXYS STARÝ AND SCHLINGER, NEW GENUS

This new genus is most closely related to *Trioxys* HALIDAY. It is a quite specialized genus, where the original paired prongs have fused into a unique median prong. Some indication of such fusion is known in certain *Trioxys* species, i.e. in *T. (T.) hincksi* MACKAUER, but in such cases the prongs are always distinctly paired.

Description: Head transverse. Occiput margined. Eyes large. Notaulices distinct at the ascendent part of mesoscutum. Propodeum areolated. Pterostigma triangular, longer than metacarp. Radial vein distinctly developed, otherwise venation reduced behind basal vein towards wing-apex. Hind wing without distinct, complete basal cell. Tergite 1 with primary tubercles. Abdomen of female lanceolate. Ovipositor sheaths and ovipositor curved downwards. Last sternite of female with one upwardly curved prong.

Type Species: Bioxys japonicus STARÝ and SCHLINGER, new species, by present designation.

Bioxys japonicus STARÝ and SCHLINGER, new species

Description – Female: Head transverse, smooth, shiny, sparsely haired, wider than thorax at tegulae. Occiput margined. Temple narrower than transverse eye-diameter (7:10). Gena as wide as $^1/_4$ of longitudinal eye-diameter. Eyes large, widely oval, convergent towards clypeus, with sparse short hairs. Interocular line $^1/_2$ longer than transfacial line (15:10), somewhat shorter than facial line (15:17). Clypeus with 6 long hairs. Tentorio-ocular line a little shorter than half on intertentorial line (2:5). Antennae broken. Socket-ocular line shorter than socket-diameter (2:3).

Mesoscutum falling vertically to prothorax, smooth, shiny, with sparse long hairs along effaced notaulices and margins. Notaulices narrow, feebly crenulate at the ascendent part, effaced on the disc and their fore margin prominent so that central lobe of mesoscutum has edged appearance when seen from above. Propodeum (fig. 66) shiny, carinated, with more or less distinct central areola. Wing (fig. 172): Pterostigma triangular, more than 3 times as long as wide. Metacarp shorter than half of pterostigma. Radial vein longer than pterostigma (30:23).

Abdomen lanceolate. Tergite 1 (fig. 101) 3 times as long as wide at spiracles, slightly convex at post-spiracular part, feebly rugose, shiny, sparsely haired. Spiracular tubercles slightly prominent. Following tergites smooth, shiny, sparsely haired. Genitalia as figured (fig. 42). Last sternite with long upwardly curved prong.

Coloration: Head brown; face, clypeus, gena and mouthparts yellow. Scape and pedicel yellow (the rest broken). Mesoscutum, scutellum and metanotum brown, remainder of thorax yellow. Wings hyaline, venation brown. Legs yellow. Tergite 1, central spot on tergite 2 and apex of abdomen yellow, remainder brown.

Length of body about 1.4 mm.

Male: Antennae broken. Head brown, mouthparts yellow. Thorax brown. Legs yellow, femora and tibiae more brown. Tergite 1 and base of tergite 2 yellow, remainder brown. Otherwise as described for the female.

Holotype ♀: Japan, Kagoshima, April 11, 1961, bred from a Callipterine aphid on *Ficus* sp. (E. I. SCHLINGER).

Allotype ♂ and 3 ♂♂ paratopotypes.

Habitat: The type specimens were all collected on a single *Ficus* leaf without mature, living aphids. Altogether fourteen mummies were collected, but nine had previously emerged. The host plant was located close to the seashore in an ornamental garden. The aphid mummies were very light brown in color.

GENUS DIAERETIELLA STARÝ

Diaeretus of authors (not FÖRSTER, 1862).
Diaeretiella STARÝ, 1960, Acta Soc. Ent. Cechosl. 57:242-3.

Type Species: Aphidius rapae M'INTOSH, by original designation.

Literary Data: STARÝ, 1961, Acta Ent. Mus. Nat. Pragae 34:383-97 (revision of world species).

This genus is related to *Aphidius* NEES, but differs from the latter by its more reduced wing-venation.

Description: Head transverse, as wide as or wider than thorax at tegulae. Antennae filiform, with variable number of segments (12 to 18). Eyes of medium size. Mandibles bidentate. Notaulices developed on the fore part of mesoscutum. Propodeum distinctly areolated. Fore wing: Pterostigma triangular. Metacarp longer than width of pterostigma. Radial vein developed, not longer than $^2/_3$ of its possible length. Otherwise venation effaced beyond basal cell towards the apex except cubital cell 2 and indicated part of cubital vein. Hind wing with complete basal cell. Abdomen of female lanceolate. Ovipositor sheaths and ovipositor straight or slightly curved upwards, sparsely haired.

General Distribution: Practically cosmopolitan.

Bionomics: Parasites of aphids with pupation occurring inside parasitized aphid.

Diaeretiella rapae (M'INTOSH)

Aphidius rapae M'INTOSH, 1855, Book of the Garden, 2, p. 194 (♀).

Aphidius rapae CURTIS, 1855, in M'INTOSH's Book of the Garden, 2, p. 194.

Diaeretus chenopodii FÖRSTER: KIRCHNER, 1867, Catal. Hymen. Europae p. 125 (*nomen nudum*).

Trioxys piceus CRESSON, 1880, U.S. Dept. Agric. Ann. Rpt. for 1879, p. 260 (♀♂, USA).

Lipolexis chenopodiaphidis ASHMEAD, 1888 (1889), Proc. U. S. Nat. Mus. 11:671 (♀♂, USA).

Aphidius brassicae MARSHALL, 1896, *in* ANDRÉ, Spec. Hym. Eur. et d'Alg. 5:597-8 (♀♂, England).

Diaeretus californicus BAKER, 1909, Pomona J. Ent. 1: 25 (♀, USA).

Diaeretus nipponensis VIERECK, 1911, Proc. U. S. Nat. Mus. 40: 182. (♀♂, Japan). TAKAHASHI, 1925, Dept. Agric. Govt. Res. Inst., Formosa, Rept. 16, in 74 pp. (Taiwan, host).

(?) *Diaeretus (Aphidius) obsoletus* KURDJUMOV, 1913, Rev. Russe ent. St. Petersburg 13:25-6 (♀♂, USSR – Ukraine).

Diaeretus napus QUILIS M. P., 1931, Eos, 7:71 (♀, Spain). QUILIS M. P., 1934, Eos, 10:6-7 (♂, Czechoslovakia, Yugoslavia).

Aphidius affinis QUILIS M. P., 1931, Eos, 7:48-50 (♀♂, Spain).

Diaeretus plesiorapae BLANCHARD, 1940, Rev. Chil. Hist. Nat. Santiago 44: 45-8 (♀♂, Argentina).

Diaeretus aphidum MUKERJI and CHATERJEE, 1950, Proc. R. Ent. Soc. London (B) 19:4-6 (♀♂, Pakistan).

Literary Data: 1961, STARÝ, Acta Ent. Mus. Nat. Pragae 34:384-90 (revision).

Description – Female: Head transverse, smooth, shiny, sparsely haired, wider than thorax at tegulae. Occiput margined. Temple as wide as transverse eye-diameter. Gena as wide as $^1/_5$ to $^1/_7$ of longitudinal eye-diameter. Clypeus transverse, oval, convex, margined frontally, smooth, shiny, with about 8 to 12 hairs, separated from face by shallow groove. Tentorio-ocular line $^1/_4$ to $^1/_6$ as long as intertentorial line. Eyes prolongately oval, of medium size, strongly convergent towards clypeus. Antenna mostly 14-segmented (rarely 13 or 15), filiform, about as long as head, thorax and tergite 1 combined. F_1 and F_2 of equal length, 2.5 times as long as wide.

Thorax smooth, shiny, sparsely haired. Mesoscutum falling almost vertically to the prothorax, without covering it when viewed from side. Notaulices distinct at fore part, deep, crenulate, effaced on disc. Propodeum areolated (fig. 62); small central pentagonal areola variable in shape. Upper lateral areola with 3 to 6, lower with 2 to 3 hairs. Wing (fig. 178): Pterostigma prolongately triangular, about 4 to 5 times longer than wide. Metacarp much shorter than pterostigma, about the same length as radial vein.

Abdomen lanceolate, longer than head and thorax combined. Tergite 1 (fig. 67) of rather variable shape, usually about 3 to 3.5 times longer than wide at spiracles, slightly convex, slightly dilated towards apex, with feeble longitudinal central carina, slightly longitudinally rugose but mostly smooth on the last fourth, with shallow lateral impressions beyond spiracular tubercles, sparsely haired. Spiracular tubercles hardly visible, situated somewhat before mid-line of tergite. Genitalia as figured (fig. 27).

Coloration: Rather variable. Head black, face sometimes yellow, clypeus and mouthparts yellow to brownish-yellow. Antennae brownish-black, with a lighter ring between pedicel and flagellar segment 1, often scape, pedicel and base of flagellar segment 1 yellow. Thorax black, prothorax sometimes yellow to brownish-yellow. Wing venation brown. Tegulae brown. Legs brown to dark brown, lower part of coxae, trochanters, base of tibiae and tarsi lighter. Abdomen brown to brownish-black, tergite 1 yellow to dark brown, base of tergite 2 and suture between tergites 2 and 3 of the same color.

Length of body about 1.9 to 2.4 mm.

Male: Antenna 16 to 18-segmented. Black, mouthparts and tergite 1 brownish-yellow. Legs brownish-black. Otherwise, coloration as in female. Tergite 1 nearly parallel-sided.

General Distribution: Probably cosmopolitan.

Material Examined: (1 ♂ specimen). Japan: Gifu, Oct. 1902 on cabbage aphid, 1 ♂. (Y. NAWA, in USNM). Note: This species was not so common in the Orient as it is in Europe and North America.

Habitat: This species was not encountered by SCHLINGER in Far East Asia,

36

but judging from its collections in many holarctic areas it ranges widely in native as well as cultivated habitats.

Hosts: (Unrevised Literature Data)

Aphidae sp.: MARSHALL, 1896, 1899, on *Raphanus raphanistrum*, England.

Aphidae sp.: MUKERJI & CHATTERJEE, 1950, on *Brassica oleracea*, Pakistan – Baluchistan.

Aphidae sp.: GIBSON & CARILLOS, 1959, on *Brassica*, Mexico.

Aphidae sp.: QUILIS M. P., 1934, on *Turritis glabra*, Czechoslovakia.

Aphis sp.: QUILIS M. P., 1934, on *Aethionema saxatile*, Yugoslavia.

Aphis abbreviata PATCH: MACGILLIVRAY & SPICER, 1953, on *Solanum tuberosum*, Canada – N. Brunswick. SHANDS, et al. 1955, USA – Maine.

Aphis rumicis L.: MACGILIVRAY & SPICER, 1953, on *Chenopodium album*, Canada-N. Brunswick.

Brachycolus noxius MORDV.: KURDJUMOV, 1913, USSR – Ukraine.

Brevicoryne brassicae (L.): BARNES, 1931, England. BEARDSLEY, 1961, Hawaii. BILANOVSKI, 1938, USSR. – Ukraine. BLANCHARD, 1940, Argentina. BUCKTON, 1879, England. FERRIÈRE & VOUKASSOVITCH, 1928, Yugoslavia. FULLAWAY, 1951, Hawaii. GEORGE, 1957, England. GOURLAY, 1930, N. Zealand. GOULAY, 1930, Europe, America, Australia, N. Zealand, S. Africa. HERRICK & HUNGATE, 1911, USA – N. Y. MARSHALL, 1896, 1899, on *Brassica oleracea*, England. MELANDER & YOTHERS, 1915, 1917, USA. MEIER, 1927, USSR – Ukraine. MUESEBECK & WALKLEY, 1951, USA. NEWTON, 1934, England. NIEZABITOWSKI, 1909, Poland. PETHERBRIDGE & MELLOR, 1936, England. RIPPER, 1944, England. SACHAROV, 1914, 1915, USSR – Ukraine. SEDLAG, 1958, 1959, Germany. SILVEIRA GUIDO & RUFFINELLI, 1958, Argentina. SMITH, C. F., 1944, USA – Utah. STRICKLAND, 1916, Canada. TELENGA, 1950, USSR – Ukraine. TIMBERLAKE, 1918, Hawaii. TODD, 1957, N. Zealand. TREHERNE, 1957, Canada. VUKASOVIC, 1928, Yugoslavia.

Callaphis betulaecolens (FITCH): MACGILLIVRAY & SPICER, 1953, on *Betula* sp., Canada – N. Brunswick (doubtful record).

Euceraphis betulae (KOCH): MACGILLIVRAY & SPICER, 1953, on *Betula* sp., Canada – N. Brunswick (doubtful record).

Hayhurstia atriplicis (L.): QUILIS M. P., 1934, on *Chenopodium album*, Czechoslovakia. BAUDYS, 1940, Czech. SMITH, C. F., 1944, USA – Idaho, on *Chenopodium album*, MACGILLIVRAY & SPICER, 1953, on *Chenopodium album*, Canada – N. Brunswick.

Hyadaphis foeniculi (PASS.): MARSHALL, 1896, 1899, on *Foeniculum vulgare,* England.

Lipaphis pseudobrassicae (DAVIS): BRITTON, 1917, USA. SMITH, C. F., 1944, USA – Utah, Ohio. MUESEBECK & WALKLEY, 1951, USA. SHANDS, et al. 1955, USA – Maine.

Macrosiphon solanifolii ASHMEAD: SPENCER, 1926, England. SILVEIRA GUIDO & RUFFINELLI, 1958, Argentina.

Myzodes persicae (SULZ.): BEARDSLEY, 1961, Hawaii. FULLAWAY, 1915, Hawaii. MACGILLIVRAY & SPICER, 1953, on *Solanum tuberosum, Brassica napus,* Canada – N. Brunswick. MASON, 1922, USA – Ida. MUESE-BECK & WALKLEY, 1951, USA. SMITH, C. F., 1944, USA – Ohio. SEDLAG, 1959, Germany. SHANDS, et al., 1955, USA – Maine. TIMBERLAKE, 1918, Hawaii. WHEELER, 1923, USA.

Aphis quilisi FRESCA: QUILIS M. P., 1931, on *Brassica napus, Raphanus sativus, Pisum sativum,* Spain.

Pterochloroides persicae CHOL.: MUKERJI & CHATTERJEE, 1950, on *Prunus persica,* Pakistan – Baluchistan (doubtful record).

Rhopalosiphum maidis (FITCH): MIMEUR, 1936, Morocco (doubtful record).

Schizaphis graminum (ROND.): KURDJUMOV, 1913, USSR – Ukraine (doubtful record). BLANCHARD, 1940, Argentina. SILVEIRA GUIDO & CONDE JAHN, 1945, Uruguay.

Hosts: (Original and Revised Literary Data)

Aphidae sp.: QUILIS M. P., 1931, on *Brassica napus,* Spain.

Aphis acanthi SCHRK.: QUILIS M. P., 1931.

Brachycaudus rumexicolens PATCH: STARÝ, 1961, on *Rumex acetosella,* Czechoslovakia.

Brachycaudus sp.: STARÝ, 1961, on *Rumex acetosella,* Czechoslovakia.

Brevicoryne brassicae (L.): STARÝ, 1961, on *Alliaria officinalis,* Czechoslovakia. SCHLINGER & HALL, 1960, on *Brassica* sp., USA – California. STARÝ, 1961, on *Brassica napus,* Czechoslovakia, on *Brassica oleracea,* Czechoslovakia, USSR – Uzbekistan, Tajikistan, on *Sinapis arvensis,* Czechoslovakia.

* NEW RECORD, Japan.

Hayhurstia atriplicis (L.): STARÝ, 1961, on *Chenopodium album,* Czechoslovakia.

Lipaphis pseudobrassicae (DAVIS): SCHLINGER & HALL, 1960, on *Malcomia maritima*, USA – California.

Myzaphis beibienkoi NARZYKULOV: STARÝ, 1961, USSR – Tajikistan.

Myzodes persicae (SULZ.): SCHLINGER & HALL, 1960, on *Vinca minor*, and *Brassica oleracea* var. *capitata*, USA – California. STARÝ, 1961, on *Beta vulgaris, Solanum tuberosum*, Czechoslovakia.

Schizaphis scirpi (KITTEL): STARÝ, 1961, on *Typha angustifolia*, Czechoslovakia.

Sitobium sp.: STARÝ, 1961, on *Lolium* sp., Czechoslovakia.

Host-specificity: Apparently this species is quite polyphagous, but it especially attacks aphids of the subfamily Myzinae, and aphids which prefer cruciferaceous and chenopodiaceous plants.

GENUS DIAERETUS FÖRSTER

Diaeretus FÖRSTER 1862, Ver. Naturh. Ver. Preuss. Rheinl. 19:249

(Type species: *Aphidius leucopterus* HALIDAY).

Literary Data: STARÝ, 1960, Acta Soc. Ent. Cechosl. 57:239-40 (revision). This genus is related to *Pauesia* QUILIS by the areolation of the propodeum, but differs from the latter by characters in wing venation, effaced notaulices on mesoscutum and by characters of the female genitalia.

Description: Head strongly transverse, wider than thorax at tegulae. Occiput distinctly margined. Eyes large. Antennae 15 to 18-segmented, filiform. Mesoscutum without notaulices. Propodeum with central carina and its bifurcation distinctly developed. Fore wing: Pterostigma triangular, longer than metacarp. Radial vein distinct, longer than width of pterostigma; all venation effaced in region behind basal cell except 2nd cubital cell and slightly colored part of cubital vein. Hind wing with complete basal cell. Abdomen of female lanceolate. Ovipositor sheaths wide, comparatively short and slightly narrowed at apex, somewhat curved. Ovipositor slightly curved downwards.

General Distribution: Europe (England, Czechoslovakia, Germany, France), Japan and South Korea.

Bionomics: Parasite of aphids *(Lachnidae)*. Pupation occurs inside parasitized aphid.

Diaeretus leucopterus (HALIDAY)

Aphidius leucopterus HALIDAY 1834, Ent. Mag. 2:103 (♀♂, England). FÖRSTER, 1862, Verh. Naturh. Ver. Preuss. Rheinl. 19:249 (as genotype of *Diaeretus*

FÖRSTER). MARSHALL, 1896, in ANDRÉ, Spec. Hym. Eur. et d'Alg. 5: 597 (♀♂, England; *Aphidius*). MARSHALL, 1899, Trans. ent. Soc. London 1899:62. (♀♂, England; *Aphidius*). STARÝ, 1960, Acta Soc. Ent. Cechosl. 57:240-2. (♀♂, Czechoslovakia, host). MACKAUER, 1961, Beitr. Ent. 11:106 (notes). STARÝ, 1962, Acta Soc. Ent. Cechosl. 59:42-58 (ecol. Czechoslovakia, host).

Aphidius exspectatus GAUTIER and BONNAMOUR 1936, Bull. Soc. Linn. Lyon N. S. 5:74 (♀, France, host).

Description – Female: Head strongly transverse, smooth, shiny, sparsely haired, wider than thorax at tegulae, strongly narrowed behind eyes. Occiput distinctly margined. Temple about as wide as transverse eye-diameter. Gena somewhat wider than base of mandible or somewhat shorter than $1/3$ of longitudinal eye-diameter. Eyes large, oval, prominent, somewhat convergent towards clypeus. Interocular line $1/3$ longer than transfacial line. Facial line $1/2$ longer than transfacial line. Facial line somewhat longer than interocular line. Clypeus oval, flat, smooth, with about 5 long hairs, separated from face by broad shallow arcuate groove. Tentorio-ocular line equal to half of (or a little shorter than half of) intertentorial line. Antennae 15 to 16-segmented, filiform, slender, reaching to about half of abdomen, situated about at the level of center of eyes. Socket-ocular line equal to socket-diameter.

Mesoscutum falling comparatively vertically to prothorax, without covering it when viewed laterally (fig. 127); smooth, shiny, sparsely haired. Notaulices effaced. Propodeum (fig. 65): central carina and its bifurcation slightly prominent and hardly visible in the longitudinal part, completing wide, central, large areola that scarcely differs by its declivity from the neighboring area. Forewing (Fig. 177): Pterostigma triangular, longer than metacarp. Radial vein somewhat longer than width of pterostigma.

Abdomen lanceolate. Tergite 1 (Fig. 68) somewhat more than 3 times as long as wide at spiracles; slender, slightly impressed behind spiracular tubercles and slightly dilating towards apex; with inconspicuous central carina reaching to about half of the tergite; basal half very slightly granulate, shiny; apical half entirely smooth, strongly convex, with sparse hairs. Spiracular tubercles not prominent, situated just before the middle of the tergite. Following tergites smooth, sparsely haired. Genitalia as figured (Fig. 29): ovipositor sheaths wide, comparatively short and slightly curved, nearly bare. Anterior prong of valvula 2 long. Ovipositor slightly curved downwards.

Coloration: Head black, mandibles brownish-yellow, apices darkened. Antennae brown. Thorax black. Wings subhyaline; venation brown. Fore leg: coxa black, other parts brown to yellow, apices of femur and tibia somewhat darkened. Middle and hind legs: coxae black, trochanters, basal fourth of tibiae, sometimes also bases of femora yellow to yellowish-brown. Femora mostly dark brown. Tibiae and tarsi brown. Abdomen dark brown; apex of tergite 1 and suture between tergites 3 and 4 slightly yellowed.

Length of body about 1.7 to 2.1 mm.

Male: Antennae (16?) 17 to 18-segmented. Coloration: Head black, mandibles brown. Thorax black. Wings almost white; venation brown. Legs dark brown, bases of middle and hind tibiae brownish-yellow. Abdomen brownish-black, with lighter spot each on tergites 3 and 4. Otherwise, as described for female except for sexual differences.

General Distribution: Europe (England, Czechoslovakia, Germany, France), Japan and South Korea.

Material Examined: (23 ♂♀ specimens). Japan: Asakawa, May 27, 1961, ex. *Eulachnus piniformosanus* on *Pinus densiflora*, 16 ♂♀ (E. I. SCHLINGER); Kagoshima, April 11, 1961, ex. *Eulachnus piniformosanus* on *Pinus densiflora*, 3 ♂♀ (E. I. SCHLINGER); Fukuoka, April 2, 1961, ex. *Eulachnus piniformosanus* on *Pinus densiflora*, 2 ♂♀ (E. I. SCHLINGER). South Korea: Seoul, May 3, 1961, ex. *Eulachnus piniformosanus* on *Pinus* sp., 2 ♂♀ (E. I. SCHLINGER).

Type: Probably lost (see MACKAUER, 1961).

Neotype ♀: Czechoslovakia, ΔZobor near Nitra, October 13, 1958, bred from "*Schizolachnus pineti*" on *Pinus nigra*, edge of wood, heath, (P. STARÝ).

Habitat: In Far East Asia this species was only encountered in pine forests.

Hosts: (Unrevised Literary Data).

　Aphidae sp.: GAUTIER & BONNAMOUR, 1936, on *Pinus*, France.

Hosts: (Original and Revised Literary Data)

★ *Eulachnus piniformosanus* TAK: on *Pinus densiflora*, Japan; on *Pinus* sp., S. Korea.

　Protolachnus spp.: STARÝ, 1960 and 1962, on *Pinus* sp., on *Pinus nigra*, Czechoslovakia.

Host-specificity: Oligophagous parasite of lachnid aphids such as *Eulachnus* and *Protolachnus*, but seemingly is specific to *Eulachnus* in Far East Asia.

Note: The mummified specimens of *Protolachnus* sp. and *Eulachnus pini-formosanus* are brownish-yellow.

GENUS EPHEDRUS HALIDAY

Ephedrus HALIDAY 1833, Ent. Mag. 1: 485. (Type species: *Bracon plagiator* NEES).

Elassus WESMAEL 1835, Nouv. Mém. Acad. Sci. Bruxelles 19:248. (Type species: *Aphidius parcicornis* NEES).

Subgenera:

1. *Ephedrus* s. str.

   Ephedrus HALIDAY subgenus *Ephedrus* s. str., STARÝ, 1958, Acta Faun. Ent. Mus. Nat. Pragae 3:66-7. (Type species: *Ephedrus (Ephedrus) plagiator* (NEES)).

2. *Lysephedrus* STARÝ

   Ephedrus HALIDAY subgenus *Lysephedrus* STARÝ, 1958, Acta Faun. Ent. Nat. Pragae 3:64. (Type species: *Ephedrus (Lysephedrus) validus* (HALIDAY), by original designation).

Literary Data: MARSHALL, 1896, in ANDRÉ, Spec. Hym. Eur. et d'Alg. 5:541-2. MARSHALL, 1899, Trans. ent. Soc. London, 1899: 20-1. SZÉPLIGETI, 1904, in WYTSMAN, Genera Insectorum, 22:183. SCHMIEDEKNECHT, 1930, Hym. Mitteleur., II. Aufl., p. 340. STELFOX, 1941, Proc. R. Irish Acad. 46(B): 128-31. (rev. of Irish spp.). SMITH, C. F., 1944, Ohio State Univ. Contr. Zool. Ent. 6:15-16 (rev. of Nearctic spp.). GRANGER, 1949, Mém. Inst. Sci. Madag. Tananarive 2(A), 1949:411. IVANOVA-KASAS, 1956, Rev. d'Ent. de l'Urss 35:245-61 (embry. and postembry. development). STARÝ, 1958, Acta Faun. Ent. Mus. Nat. Pragae 3:61-2 (rev.of European spp.). LUZHETZKI, 1960, Par. tlej Uzbekistana, pp. 110-11 (Centr. Asia). IVANOVA-KASAS, 1961, Otsh. po sravn. anat. perepontshatokrylych, in 265 pp. (embryology). STARÝ, 1962, Opusc. entomol. 27:87-98 (rev. of European spp.).

Ephedrus is similar to *Toxares* HALIDAY, but differs from latter by shape of ovipositor sheaths and in possessing 11-segmented antennae in both sexes.

Description: Head transverse, as wide as or wider than thorax at tegulae. Antennae 11-segmented in both sexes. Notaulices at least partly distinct. Venation of wing nearly complete (of subfamily rank). Fore wing: Pterostigma triangular, more or less prolonged at both ends. Pterostigmal cell complete. Radial vein reaching wing margin. Median vein complete and three radial cells developed. Cubital vein distinct for its greatest part. Median cell complete. Hind wing with complete basal cell. Propodeum either distinctly areolated, with areolae smooth or slightly rugose along carinae, or rugose, with carinae more or less distinct. Abdomen lanceolate in females and rounded in males. Tergite 1 either slender and long or nearly square, with various sculpturing (coarsely rugose to nearly smooth). Ovipositor sheaths either slender and long, with scattered hairs and nearly truncate or rounded at apex, or wide, relatively quickly narrowing towards apex, densely pubescent throughout and bluntly point at apex.

General distribution: Palearctic, Nearctic, Ethiopian and Oriental regions.

Bionomics of species: Parasites of aphids, and pupation occurs inside host. Parasitized aphids (mummies) are dull or shiny black, or rarely dark brown.

Key to the subgenera and species of *Ephedrus* (♀♀)

1   Propodeum coarsely, irregularly and deeply rugose. Ovipositor sheaths densely pubescent (Subg. *Lysephedrus* STARÝ) *E.* (L.) *validus* (HALIDAY)

     Propodeum smooth, regularly areolated, area around carinae slightly sculptured. Ovipositor sheaths with scattered hairs (Subg. *Ephedrus* s. str.) 2

2(1) Radial abscissa 2 distinctly shorter than interradial vein 1 3

     Radial abscissa 2 equal to or longer than interradial vein 1 4

3(2) Mesoscutum with distinct, deep fovea in the confluent point of notaulices. Tergite 1 nearly square (Fig. 141)   *E. (E.) persicae* FROGGATT

     Mesoscutum without deep fovea in the confluent point of notaulices. Tergite 1, $2^1/_2$ times as long as wide at spiracles (Fig. 144)

                 E. (E.) orientalis STARÝ and SCHLINGER n. sp.

4(2) Radial abscissa 2 equal to interradial vein 1 5

     Radial abscissa 2 at least a little or distinctly longer than interradial vein 1   *E. (E.) plagiator* (NEES)

5(4) F_1 of antenna about 6 or 7 times as long as wide, $^1/_2$ longer than F_2. F_1 brownish-yellow to yellow. Radial abscissa 1 longer than width of pterostigma (Fig. 157)   *E. (E.) lacertosus* (HALIDAY)

     F_1 of antenna about 4 times as long as wide, stout, $^1/_5$ to $^1/_6$ longer than F_1. F_2 black, brown at base. Radial abscissa 1 shorter than width of pterostigma (Fig. 158)   *E. (E.) campestris* STARÝ

SUBGENUS *Ephedrus* HALIDAY

Head transverse, as wide as or wider than thorax. Antennae 11-segmented. Mesoscutum smooth, sparsely haired. Notaulices at least partly distinct. Propodeum areolated, discs of areolae smooth, sometimes slightly rugose along carinae. Abdomen lanceolate, sparsely haired. Tergite 1 rugose or partly smooth. Ovipositor sheaths long and narrow, gradually and evenly narrowing towards apex; sheaths with scattered hairs.

General distribution: Palearctic, Nearctic, Ethiopian and Oriental regions.

Ephedrus (E.) campestris STARÝ

Ephedrus (Ephedrus) campestris STARÝ, 1962, Opusc. entomol. 27:87-91 (♂♀, Czechoslovakia, European part of USSR, hosts).

Diagnosis: This species is closely related to *E. plagiator* (NEES), but differs from the latter by the relative lengths of the interradial vein 1 and radial abscissa 2, which are equal in *E. campestris*, while interradial vein 1 is always a little and usually distinctly shorter than radial abscissa 2 in *E. plagiator*, and by the shape of tergite 1. It parasitizes *Dactynotus* and *Macrosiphoniella* species in Europe, and also *Megoura viciae* in Far East Asia.

Description – Female: Head transverse, smooth, shiny, sparsely haired, roundly narrowed behind eyes, as wide as thorax at tegulae. Temple as wide as transverse eye-diameter. Gena as wide as $^1/_3$ of longitudinal eye-diameter. Eyes small, oval, slightly prominent, sparsely and shortly haired, slightly convergent towards clypeus. Clypeus slightly convex, with 7 to 9 long hairs. Intertentorial line $^1/_3$ longer than tentorio-ocular line. Antennae 11-segmented filiform, reaching apex of tergite 1. F_1 4 times as long as wide at apex and $^1/_5$ longer than the F_2. (Fig. 150). Preapical and apical segments distinctly separated. Socket-ocular line equal to half of socket-diameter.

Mesoscutum at base of central lobe, along its margins and along effaced notaulices with sparse, long hairs. Notaulices distinct at the ascendent part, wide, deeply crenulate, effaced on disc. Propodeum (Fig. 11) areolated. Central areola wide, pentagonal. Discs of areolae smooth, shiny, more or less rugose near carinae. Upper areola with 6 to 8, lower with 3 to 5 hairs. Wing (Fig. 158): Pterostigma triangular, 4 times as long as wide. Radial abscissa 1 about as long as $^1/_2$ the pterostigma width. Interradial vein 1 and radial abscissa 2 of equal length.

Abdomen lanceolate. Tergite 1 (Fig. 147) somewhat more than twice as long as wide at spiracles, nearly parallel-sided, a little wider behind than at spiracles; with 1 central and 2 lateral, strongly prominent, longitudinal carinae; with deep lateral impressions on apical third; surface nearly smooth, with slight rugosities, sparsely haired. Spiracular tubercles hardly visible, situated at end of first third of tergite. Following tergites smooth, sparsely haired. Genitalia as figured (Fig. 32).

Coloration: Head black, mouthparts brown. Antennae black, base of F_1 yellow. Thorax black. Wings almost hyaline, venation dark brown. Tegulae brown. Legs brownish-black, basal parts of tibiae brown. Abdomen brownish-black, tergite 1 sometimes lighter colored.

Length of body about 2.2 to 3.1 mm.

Male: Similar to female except for sexual differences.

General distribution: Europe (Czechoslovakia, European part of USSR) and South Korea.

New material examined (40 ♂♀ specimens). South Korea: Seoul, May 9, 1961, ex. *Megoura viciae* on *Vicia* sp., 10♂♀, (E. I. SCHLINGER). Seoul, May 9, 1961, ex. *Macrosiphoniella yomogifoliae* on *Artemisia manischmidtiana*, 15 ♂♀, (E. I. SCHLINGER). Seoul, May 9, 1961, ex. *Macrosiphoniella sanborni* on *Chrysanthemum* sp., 15 ♂♀, (E. I. SCHLINGER).

Holotype ♀: Czechoslovakia, Somotor, June 7, 1961, *Dactynotus jaceae* on *Centaurea stoebe*, steppe, (P. STARÝ), deposited in P. Starý collection. *Allotype* is topotypical.

Habitat: In Europe this species occurs in steppe or to a lesser degree in wood-steppe habitats. The only encounter made with this species in Far East Asia was in the Forestry Botanical Garden in Seoul, South Korea.

Hosts: (Original and revised literature data).

Dactynotus cichorii (KOCH): STARÝ, 1962, on *Leontodon hispidus*, and *Cichorium intybus* L., Czechoslovakia.

Dactynotus muralis (BCKT.): STARÝ, 1962, on *Lactuca quercina*, Czechoslovakia.

Dactynotus obscurus (KOCH): STARÝ, 1962, on *Hieracium* sp., Czechoslovakia.

Dactynotus picridis (F.): STARÝ, 1962, on *Picris hieracioides*, Czechoslovakia.

Dactynotus sp.: STARÝ, 1962, on *Crepis biennis*, Czechoslovakia.

Macrosiphoniella absinthii (L.): STARÝ, 1962, on *Artemisia absinthium*, Czechoslovakia and European part of USSR.

Macrosiphoniella millefolii (DEG.): STARÝ, 1962, on *Achillea nobilis* and *Achillea millefolium*, Czechoslovakia.

Macrosiphoniella sanborni (GILL): On *Chrysanthemum* sp., South Korea.

Macrosiphoniella yomogifoliae TAKAH.: On *Artemisia manischmidtiana*, South Korea.

Megoura viciae BCKT.: On *Vicia* sp., South Korea.

Dactynotus jaceae (L.): STARÝ, 1962, on *Centaurea stoebe* and *Centaurea scabiosa*, Czechoslovakia.

Dactynotus aeneus HRL.: STARÝ, 1962, on *Carduus nutans*, Czechoslovakia.

Host-specificity: Oligophagous with *Dactynotus*, and *Macrosiphoniella* species being the main hosts in Europe. In Far East Asia it is a parasite of *Macrosiphoniella* species and *Megoura viciae*. Observations in Seoul, South Korea indicated that *Macrosiphoniella* species were greatly preferred as hosts over *Megoura viciae* notwithstanding the number of reared specimens.

Ephedrus (E.) lacertosus (HALIDAY)

Aphidius (Ephedrus) lacertosus HALIDAY, 1833, Ent. Mag. 1:486 (♀♂, England). MARSHALL, 1896, in ANDRÉ, Spec. Hym. Eur. et d'Alg. 5:543 (♀♂, England, host; *Ephedrus*). MARSHALL, 1899, Trans. Ent. Soc. London 1899:21-2 (♀♂, England, host). STELFOX, 1941, Proc. R. Irish Acad. 46 (B): 136-7 (♀♂, Ireland). STELFOX, 1944, Ent. mon. Mag. 80:236 (Scotland). STARÝ, 1958, Acta Faun. Ent. Mus. Nat. Pragae 3:70-2 (♀♂, Czechoslovakia, host). MACKAUER, 1961, Entomophaga, 11:99 (notes on the type). STARÝ, 1962, Opusc. entomol. 27:92-3 (ecol., Czechoslovakia, hosts).

Diagnosis: The very long F_1, the shape of wing-venation with the long and slender radial cell easily distinguishes this species from its congeners.

Description – Female: Head transverse, smooth, shiny, sparsely haired, wider than thorax at tegulae. Temple a little narrower than transverse eye-diameter. Gena as wide as $1/3$ of longitudinal eye-diameter. Eyes of medium size, oval, sparsely and shortly haired, slightly convergent towards clypeus. Interocular line about $1/5$ longer than transfacial line and about $1/6$ shorter than facial line. Clypeus with about 6 hairs. Tentorio-ocular line $1/2$ of inter-tentorial line or somewhat longer. Antennae 11-segmented, filiform, reaching to about $1/2$ of abdomen. F_1 very long, 6 to 7 times as long as wide, about $1/2$ longer than F_2 (Fig. 148). Socket-ocular line less than $1/2$ of socket-diameter.

Mesoscutum sparsely, but long haired. Notaulices wide, deep, rugose on the anterior half, smooth behind, and more or less distinct on the disc.

Propodeum areolated (Fig. 105) with central pentagonal areola. Wing (Fig. 157): Pterostigma very long and slender. Radial abscissa 1 longer than width of pterostigma, nearly perpendicular to it. Interradial vein 1 and radial abscissa 2 of equal length. Median abscissa 2 twice as long as interradial vein 1.

Abdomen lanceolate. Tergite 1 (Fig. 145) slender, 2.5 times or more longer than wide at spiracles, nearly parallel-sided, sparsely haired, with more or less distinct central and 2 lateral longitudinal carinae, more or less rugose, with slight lateral impressions before the apex. Spiracular tubercles not prominent. Following tergites smooth, shiny, sparsely haired. Genitalia as figured (Fig. 36).

Coloration: Variable, brown to brownish-black. Head brown, mouthparts yellow. Antennae brown except scape, pedicel and the greater part of F_1 yellow. Thorax brown, prothorax somewhat lighter. Wings hyaline, venation brown. Legs yellow, apices of tarsi darker. Abdomen brown, tergite 1 and apex more yellow. Ovipositor sheaths yellow with darker apex.

Length of body about 2.1 to 2.9 mm.

Male: Similar to female except for sexual differences.

General distribution: Europe and Taiwan (Formosa).

New material examined (2 ♀♂ specimens). Taiwan: Wulai, March 21, 1961, ex. *Myzus woodwardiae* on *Woodwardia* sp., 1 ♀, 1 ♂, (E. I. SCHLINGER).

Holotype ♂: England, B. M. Type Hym. 3. x. 71.

Habitat: This species was only encountered once in Far East Asia and when found was extremely abundant along road sides on *Woodwardia* sp. (ferns) in a moist, low mountain region near the village of Wulai. Unfortunately, of the 75 mummies collected all but 2 were hyperparasitized.

Hosts: (Unrevised literature data).

Aphidae sp.: MARSHALL, 1896, on *Ervum hirsutum*, England.

Amphorophora rubi KALT: NARZYKULOV & ATAEVA, 1961, on *Rubus caesius*, USSR – Tajikistan.

Aphis pomi DEG.: WIACKOWSKI & WIACKOWSKA, 1961, Poland. TELENGA, 1950, USSR – Ukraine.

Cryptomyzus ribis L.: DOBROVLIANSKY, 1913, USSR – Ukraine. MEIER, 1927.

Dysaphis pyri FONSC.: SAVARY, 1953, Switzerland.

Hyalopterus pruni DEG.: WIACKOWSKI & WIACKOWSKA, 1961, Poland. DILL, 1937, on *Prunus* and *Phragmites*, Switzerland.

Macrosiphoniella absinthii L.: FERRIÈRE & VOUKASSOVITCH, 1928, Switzerland.

Myzus cerasi F.: MARSHALL, 1896, England.

Rhopalosiphon oxyacanthae (SCHRK.): DOBROVLIANSKY, 1913, USSR – European part. MEIER, 1927.

Hosts: (Original and revised literature data).

Macrosiphum rosae (L.): STARÝ, 1958, 1962, on *Rosa* sp., Czechoslovakia.

*Myzus woodwardiae TAKAH.: On *Woodwardia* sp., Taiwan.

Rhopalosiphoninus sp. nr. *tulipellae* THEO.: STARÝ, 1962, on *Oxalis* sp., Czechoslovakia.

Ephedrus (E.) persicae FROGGATT

Ephedrus persicae FROGGATT, 1904, Agric. Gaz. Sydney, 15:611–612, fig. 6 (♀, New South Wales, Australia). See MACKAUER, 1963.

Ephedrus nitidus GAHAN, 1917, Proc. U. S. Nat. Mus. 53:195 (♀). MCLEOD, 1938, Rep. ent. Soc. Ont. 68 (1937):44-8 (Canada, hosts). SMITH, 1944, Ohio State Univ. Contr. Zoo. Ent. 6:21 (♀, USA: New Jersey, Ohio, Calif., Canada). MUESEBECK & WALKLEY, 1951, in U. S. Dept. Agric. Monogr. 2:91 (USA – Ont. to Va., Ohio, Calif., hosts). PROVERBS, 1954, Proc. Ent. Soc. B. C. 51: in 23-30 (Brit. Col., effects of insecticides, hosts). KROMBEIN, 1958, U. S. Dept. Agric. Monogr. 2, 1st Suppl., p. 18 (USA and B.C. Can., host). SCHLINGER & HALL, 1960, Ann. Ent. Soc. Amer. 53: 410 (USA – Calif., diapause, hosts). GILMORE, 1960, J. Econ. Ent. 53(4): 659-61 (U.S.A. – Oregon, host). STARÝ, 1963a, Boll. Ent. Agr. Labor Portici 21:207 (hosts).

Ephedrus vidali QUILIS M.P., 1934, Eos, 7:72-4 (♀♂, Spain). STARÝ, 1958, Acta Faun. Ent. Mus. Nat. Pragae 3:81-2 (notes). STARÝ, 1962, Opusc. entomol. 27:96 (synonymy, notes on the type).

Ephedrus pulchellus STELFOX, 1941, Proc. R. Irish Acad. 46B: 139 (♀, Ireland). BEIRNE, 1942, Ent. Mon. Mag. 78:283 (Ireland, host). STELFOX, 1944, Ent. mon. Mag. 80:236 (Ireland, host). STARÝ, 1958, Acta Faun. Ent. Mus. Nat. Pragae 3:79-81 (♀♂, Czechoslovakia, hosts). MACKAUER, 1959, Beitr. Ent. 9:866-7 (Israel, hosts). STARÝ, 1962, Acta Soc. Ent. Cechosl. 59: in 42-58 (ecol., hosts in Czechoslovakia). STARÝ, 1962, Opusc. entomol. 27: in 87-98 (ecol., hosts in Czechoslovakia). STARÝ, 1962, Entomophaga, 7:91 – 100 (bion. ecol., hosts).

Ephedrus interstitialis WATANABE, 1941, Ins. Mats. 15:139-40 (♀, Japan, host).

Diagnosis: This species belongs to that group of species which is characterized by having the radial abscissa 2 shorter than the interradial vein 1. It differs from its relatives by the distinct, well-developed fovea in the confluent point of the notaulices.

Description – Female : Head transverse, rounded, smooth, shiny, sparsely haired, as wide as thorax. Temple equal to transverse eye-diameter. Gena as long as about $^1/_3$ to $^1/_4$ of longitudinal eye-diameter. Eyes of medium size, oval, with sparse short hairs, somewhat convergent near clypeus. Interocular line $^1/_3$ to $^1/_4$ longer than transfacial line and about $^1/_3$ to $^1/_4$ shorter than facial line. Clypeus with about 10 to 15 hairs. Tentorio-ocular line a little shorter than $^1/_2$ of intertentorial line. Antenna 11-segmented, filiform, reaching to apex of tergite 1. F_1 about 5 times as long as wide, longer than F_2. Socket-ocular line equal to half of socket-diameter.

Mesoscutum with sparse long hairs. Notaulices rugos at the ascendent part and more or less visible throughout, with deep fovea in the confluent point before praescutellar groove. Propodeum (Fig. 109) areolated, with central pentagonal areola. Discs of areolae nearly smooth or with some irregular rugosities. Wing (Fig. 159): Pterostigma 6 to 7 times as long as wide. Radial abscissa 1 about 1/2 as long as radial abscissa 2. Interradial vein 1 half again as long as radial abscissa 2.

Abdomen lanceolate. Tergite 1 (Fig. 141) nearly square, convex, with variable sculpture from rugose to nearly smooth with only the carinae distinct, with one central and two lateral more or less distinct carinae. Spiracular tubercles obtusely prominent laterally. Following tergites smooth, shiny, sparsely haired. Genitalia as figured (Fig. 35).

Coloration : Black. Base of antennae (scape, pedicel and F_1) and mouthparts more or less yellowish-brown. Legs yellow to yellowish-brown, sometimes coxae, femora and tarsi darker. Tergite 1 more or less yellowish-brown to brown; sometimes center of abdomen yellowish-brown, remainder brown.

Length of body about 1.3 to 2.2 mm.

Male : Apical and preapical antennal segments distinctly separated. Otherwise as described for female except for sexual differences.

General distribution : Australian region, Palearctic region (Europe, Asia Minor, Central Asia, USSR (Primorye), South Korea and Japan), Oriental region, (Hong Kong and Taiwan), Nearctic region (USA and Canada).

New material examined (134 ♀♀, and no ♂ specimens). Japan: Fukuoka, April 18, 1961, ex. *Myzus momonis* on *Prunus* sp., 10 ♀♀ (E. I. SCHLINGER);

Fukuoka, April 18, 1961, ex. *Capitophorus* sp. on *Elaeagnus* sp., 10 ♀♀ (E. I. SCHLINGER); Fukuoka, April 18, 1961, ex. *Vesiculaphis caricis* on *Rhododendron* sp., 5 ♀♀ (E. I. SCHLINGER); Ibusuki, April 17, 1961, ex. *Myzus* sp. on *Prunus* sp., 4 ♀♀ (E. I. SCHLINGER); Tokyo, May 7, 1961, ex. *Myzus momonis* on *Prunus* sp., 6 ♀♀ (E. I. SCHLINGER); Tokyo, April 25, 1961, ex. *Myzus persicae* on *Weigela coraensis*, 1 ♀ (E. I. SCHLINGER); Tokyo, May 13, 1961, ex. *Myzus malsuctus* on *Chaenomelles japonica*, 3 ♀♀ (E. I. SCHLINGER). USSR: July 15, 1961, Primorye, Artjom, near Vladivostok, 3 ♀♀ (TRIAPICYN). Taiwan: Wushe, March 12, 1961, ex. *Myzus momonis* on *Prunus persica*, 41 ♀♀ (E. I. SCHLINGER); Taipei, March 20, 1961, ex. *Myzus persicae* on *Hibiscus* sp., 1 ♀ (E. I. SCHLINGER); Taipei, March 3-4, 1961, ex. *Aphis gossypii* on *Chrysanthemum* sp., 6 ♀♀ (E. I. SCHLINGER): Taipei, March 6, 1961, ex. *Aphis gossypii* on *Hibiscus* sp., 1 ♀ (E. I. SCHLINGER); Taipei, March 20, 1961, ex. *Agrioaphis viridis* on *Ulnus* sp., 3 ♀♀ (E. I. SCHLINGER); Taipei, March 24, 1961, ex. *Aphis spiraecola* on *Spiraea cantonensis*, 1 ♀ (E. I. SCHLINGER). South Korea: Chejudo Island, May 6, 1961, ex *Aphis gossypii*, 2 ♀♀ (E. I. SCHLINGER); Chejudo Island, May 6, 1961, ex. *Myzus persicae* on *Prunus* sp. 29 ♀♀ (E. J, SCHLINGER). Hong Kong: Kowloon, March 1, 1961, ex. *Myzus persicae* on *Viola* sp., 1 ♀ (E. I. SCHLINGER); Sha Tin, Feb. 27, 1961, ex. *Myzus persicae* on *Brassica* sp., 3 ♀♀ (E. I. SCHLINGER).

Type specimens:

Ephedrus persicae FROGGATT, ♀, New South Wales, Australia (location unknown).

Ephedrus nitidus GAHAN: ♀, USA – NB, NJ (deposited in USNM No. 20373).

Ephedrus vidali QUILIS M. P.; ♀, Spain, Azuebar, 11.v. 1930, sobre higuero (deposited in coll. Estación de Fitop. agrícola, Burjasot, Valencia).

Ephedrus pulchellus STELFOX: ♀, Ireland, 17. VI. 1937, Lough Neagh, co. Armagh. (deposited in coll. Stelfox).

Ephedrus interstitialis WATANABE: ♀, Japan, Sapporo, 10. VI. 1939, *Myzus mumecola*, coll. Watanabe (deposited in coll. of Hokkaido University, Sapporo).

Habitat: In Europe this species occurs as the typical species of the forest-type habitats while it is absent in the fields. In Central Asia it occurs mainly in oases. In Far East Asia it is typically an arboreal species usually associated with aphids in curled leaves of *Prunus* species, but it also can parasitize suitable aphid hosts in less favorable habitats. In California, USA, it is found in various undisturbed habitats featuring *Myzus persicae*.

Hosts: (Unrevised literature data).

Aphis fabae SCOP.: MACKAUER, 1959, on *Philadelphus coronarius*, Germany.

Aphis ruborum (BOERNER): MACKAUER, 1959, Israel.

Dysaphis mali (FERR.): BEIRNE, 1942, on *Malus silvestris*, Ireland.

Dysaphis plantaginea (PASS.): MACKAUER, 1959, on *Malus silvestris*, Germany.

Macrosiphum solanifolii ASHM.: SMITH, 1944, USA – Ohio.

Myzus cerasi F.: KROMBEIN, 1958, USA and Canada – B.C. GILMORE, 1960, USA – Oregon.

Myzus momonis (MATS.): WATANABE, 1941, on *Prunus donarium* var. *sachaliensis*, Japan.

Myzus mumecola MATS.: WATANABE, 1941, on *Prunus americana*, Japan.

Myzus persicae SULZ.: MCLEOD, 1938, on tobacco, tomatoes, USA. SMITH, 1944, USA – Calif. PROVERBS, 1954, Canada – B.C.

Hosts: (Original and revised literary data).

Agrioaphis viridis TAKAH.: On *Ulnus* sp., Taiwan.

Allocotaphis quaestionis (BOERNER): STARÝ, 1962, on *Malus silvestris*, Czechoslovakia.

Aphis fabae SCOP.: STARÝ, 1962, on *Euonymus europaea*, Czechoslovakia.

Aphis gossypii GLOV.: SCHLINGER & HALL, 1960, on *Citrus sinensis*, California and South Korea. On *Chrysanthemum* sp., and *Hibiscus* sp., Taiwan.

Aphis medicaginis KOCH: USSR – Tajikistan.

Aphis pomi DEG.: USSR – Tajikistan.

Aphis spiraecola PATCH: On *Spiraea cantonensis*, Taiwan.

Aphis sp.: SCHLINGER & HALL, 1960, on *Foeniculum vulgare*, California.

Aphidae sp.: On *Populus tremula*, Poland.

Brachycaudus cardui (L.): STARÝ, 1962, on *Prunus spinosa* and *Prunus domestica*, Czechoslovakia.

Brachycaudus helichrysi KALT.: STARÝ, 1962, on *Anthemis* sp., Czechoslovakia.

Brachycaudus sp.: STARÝ, 1962, on *Melandrium* sp., Czechoslovakia.

Brachycaudus sp.: STARÝ, 1962, on *Prunus persica*, Czechoslovakia.

Capitophorus sp.: On *Elaeagnus* sp., Japan.

Dysaphis crataegi KALT.: USSR – Tajikistan.

Dysaphis devecta WALK.: STARÝ, 1958, 1962, on *Malus silvestris*, Czechoslovakia.

Dysaphis sorbi (KALT.): STARÝ, 1962, on *Sorbus aucuparia*, Czechoslovakia.

Dysaphis sorbiarum NARZ.: USSR – Tajikistan.

Dysaphis spp.: STARÝ, 1962, on *Malus silvestris*, *Sorbus torminalis*, *Crataegus oxyacantha* and *Pyrus communis*.

Myzus cerasi F.: STARÝ, 1958, 1962, on *Prunus avium*, Czechoslovakia.

Myzus ligustri MOSL.: STARÝ, 1962, on *Ligustrum vulgare*, Czechoslovakia.

★*Myzus malsuctus* MATS.: On *Chaenomelles japonica*, Japan.

★*Myzus momonis* MATS.: On *Prunus persica*, Taiwan. On *Prunus* sp., Taiwan and Japan.

★*Myzus persicae* SULZ.: SCHLINGER & HALL, 1960, on *Vinca minor*, California, on *Viburnum suspensum*, California. On *Brassica* sp., and *Viola* sp., Hong Kong. On *Hibiscus* sp., Taiwan. On *Prunus* sp., South Korea. On *Weigela coraensis*, Japan.

★*Myzus* sp.: On *Prunus* sp., Japan.

Phorodon humuli (SCHRK.): STARÝ, 1962, on *Humulus lupulus* and *Prunus domestica*, Czechoslovakia.

Rhopalosiphum padi (L.): STARÝ, 1962, on *Padus racemosa*, Czechoslovakia.

Roepkea marchali (BOERNER): STARÝ, 1962, on *Prunus mahaleb*, Czechoslovakia.

★*Vesiculaphis caricis* FULLA.: On *Rhododendron* sp., Japan.

Host-specificity: The degree of host-specificity throughout the known area of the parasite's distribution appears to be quite wide. Altogether 30 species are recorded as hosts of *E. persicae* (as *nitidus)*. See recent paper by MACKAUER (1963) on host list, etc.

In Europe the anuraphidine *(Roepkea, Allocotaphis, Dysaphis, Brachycaudus)* and myzine groups *(Myzus)* of aphids are obviously preferred, while to a lesser degree the Aphidina *(Rhopalosiphum, Aphis)* are utilized. In the Nearctic region, as far as can be concluded at present the myzine and aphidine groups are preferred. In Far East Asia, *E. persicae* was found to be almost exclusively a parasite of *Myzus* species, and was only commonly encountered in cryptic, leaf-curled conditions.

Apparently the host-specificity is determined by the arboricolous type of habitat, which is obviously preferred. Presumably the original group of aphids to which *E. persicae* has become adapted to is the anuraphidine group, in which group it parasitizes the greatest number of genera. However, it occurs as a parasite of this group primarily in Europe. The original distribution of *E. persicae* appears to be European since it is very common and occurs there as an obligatory biparental species. It later extended its range of hosts into other aphid groups (i.e. the Myzinae and Aphidinae) which have more or less different phylogenetic relationships, but a similar mode of life.

Sex. *E. persicae* occurs commonly in Europe as an obligatorily biparental species. A similar situation seems to occur in Central Asia. In the Nearctic region, as far as it is known, males are recorded only from Canada. In Far East Asia, judging from our rather large number of specimens, *E. persicae* is an obligatorily uniparental species. Also, the uniparental species is the only one that is known to occur in California, according to SCHLINGER (unpublished notes).

It seems possible that there exists both a holarctic biparental *E. persicae*, and at least a Far East Asian and a Californian, uniparental form. Whether these actually represent one polytypic species, several sibling species, or several biological races will not be known until further biological studies are undertaken. One clue to a possible distinction between the uniparental and biparental forms is the different host aphids attacked. The uniparental form seems to prefer myzine aphids while the biparental form apparently prefers anuraphidine aphids.

Note: Diapause. A certain part of the progeny of *E. persicae* have a larval diapause at the end of spring and at the beginning of summer in Europe. The mummified aphids, including the diapause-cocoons, are unusually large, globular in shape, shiny, strongly mummified and the segmentation of the host's abdomen is unrecognizable. They are attached weakly to the leaf surface (see STARÝ 1962). Diapause in this species was first reported from the Nearctic region (California), by SCHLINGER & HALL (1960).

Ephedrus (E.) orientalis STARÝ and SCHLINGER, new species

Diagnosis: This species is related to *E. persicae* FROGGATT by the wing venation, but differs from the latter in the absence of foveae in the confluent point of the notaulices, and by the shape of tergite 1.

Description – Female: Head transverse, smooth, shiny, sparsely haired, somewhat wider than thorax at tegulae. Occiput margined. Temple equal or a little shorter than transverse eye-diameter. Gena a little shorter than $1/_3$

of longitudinal eye diameter (4:15). Eyes of medium size, oval, with sparse short hairs, slightly prominent, slightly convergent towards clypeus. Interocular line $1/_3$ longer than transfacial line, somewhat shorter than facial line, (about 18:20). Clypeus transverse, with about 7 to 10 long hairs. Tentorio-ocular line a little longer than $1/_3$ of intertentorial line. Antenna 11-segmented, filiform, as long as head and thorax combined. F_1 more than 3 times as long as wide at apex, somewhat longer than F_2 (Fig. 151). F_1 with 2, F_2 with 3, F_3 with 4 rhinaria. Preapical and apical segments distinctly separated. Socket-diameter twice as long as socket-ocular line.

Mesoscutum raised above prothorax, without covering it when viewed laterally, smooth, shiny, with long, sparse hairs at the base of central lobe, along margins and along effaced notaulices on the disc. Notaulices deep, crenulate on anterior part, effaced on disc. Propodeum (Fig. 110) areolated, with narrow central areola. Upper areola with 7, lower with 3 hairs. Wing similar to fig. 159, except: pterostigma 4.5 times as long as wide; radial abscissa 1 equal to $1/_2$ pterostigma-width; radial abscissa 2, 3 to 4 times as long as abscissa 1, distinctly shorter than interradial vein 1.

Abdomen lanceolate. Tergite 1 (Fig. 144) 2.5 times as long as wide at spiracles, nearly parallel-sided, a little wider at apex than at spiracles, with strong, prominent, bifurcate, central longitudinal carina and less distinct lateral carinae; surface slightly rugose, sparsely haired; with deep impression on the apical third. Following tergites smooth, shiny, sparsely haired. Genitalia as figured (Fig. 33).

Coloration: Head brownish-black to brown; mouthparts brownish-yellow. Antennae brown, pedicel and base of F_1 yellow. Thorax brownish-black. Wings hyaline, venation brownish-yellow. Legs yellowish-brown. Tergite 1 dark brown, following tergites lighter colored.

Length of body about 2.8 mm.

Male: As described for female except for sexual differences.

General distribution: Taiwan (Formosa).

Material examined: Taiwan: Taipei (Taihoku), June 18, 1926, ex. *Amphorophora oleracea*, holotype ♀, allotype ♂, and 10 ♂♀ paratypes (TAKAHASHI).

Holotype ♀: Taiwan, Taihoku, ex. *Amphorophora oleracea*, June 18, 1926, USNM type No. 65940.

Allotype ♂: Topotypical with the holotype.

Host: Amphorophora oleracea v. D. GOOT: Taiwan.

Note: All the specimens are deposited in U. S. National Museum col-

lection except for 2 ♀♀ paratypes in the P. Starý collection, and 5 ♂♀ paratypes in the University of California at Riverside collection.

Ephedrus (E.) plagiator (NEES)

Bracon plagiator NEES, 1811, Mag. Ges. Nat. Fr. Berlin 5:17 (♀♂, Germany).
HALIDAY, 1833, Ent. Mag. 1:486 (*Aphidius*, subg. *Ephedrus*). WESTWOOD,
1840, Introd. Mod. Classif. Insects I, Synopsis p. 65 (*Ephedrus*). MARSHALL,
1896, in ANDRÉ, Spec. Hym. Eur. et d'Alg. 5:542-4 (♀♂, Germany, Italy,
Belgium, England, Spain, Netherlands, hosts, *Ephedrus*). MARSHALL, 1899,
Trans. Ent. Soc. London 1899:22-3 (♀♂, see 1896). REGNIER, 1923, Rev.
Bot. appl. Agric. colon. 3:169-85 (host). MEIER, 1927, Repts. Bur. appl.
Ent. Lenn. 3:(Eur. part of USSR, host). GAUTIER & BONNAMOUR, 1929,
Bull. Soc. Ent. Fr. 1929:92-5 (♀♂, France). SKRIPTSHINSKIJ, 1930, Rep. Appl.
Ent. Lenn. 4:(♀♂, embryology, USSR). LINDROTH, 1931, Zoll. Bidr. Upp-
sala 13:349 (Iceland, host). QUILIS M. P., 1934, Eos, 10:16-17 (Czechoslo-
vakia, host). STELFOX, 1941, Proc. R. Irish Acad. 46 B:131-5 (♀♂, Ireland).
TELENGA, 1950, Nautsh. Trudy Inst. Ent. Phytopath. AN Ukr. SSR Kiev 2:
(USSR – Ukraine, hosts). PETERSEN, 1956, Zoology of Iceland, 3, Pt. 49-50:
42-3 (Iceland, host). IVANOVA-KASAS, 1956, Revue d'Ent. de l'URSS 35
(USSR, embryology, hosts). STARÝ, 1958, Acta Faun. Ent. Mus. Nat. Pragae
3:76-9 (♀♂, Czechoslovakia, Europ. part of USSR, hosts). STARÝ, 1959,
Trans. I. Int. Conf. Insect Pathology and Biol. Control, Praha (1958): in
537-41 (ecol., hosts, Czechoslovakia). LUZHETZKI, 1959, Tez. Dokl. 4-ogo
sj. Vses. Obsth. Moscow-Leningrad, p. 82 (USSR – Uzbekistan). THUNE-
BERG, 1960, Suomen hyönt. aikakauskirja 26(1): 99 (Finland, hosts).
LUZHETZKI, 1960, Par. tlej Uzbekistana, p. 111-2 (♀♂, USSR – Uzbekistan,
hosts.) [Note: This record belongs to *E. persicae* FROGGATT]. WIACKOWSKI &
WIACKOWSKA, 1961, Bull. ent. Pologne 31: in 255-62 (Poland, hosts).
STARÝ, 1962, Acta Soc. Ent. Cechosl. 59: in 42-58 (ecol., hosts, Czechoslovakia).

Aphidius parcicornis NEES, 1834, ¦Hym. Ichn. aff. Mon. 1: 16 (♀♂, Germany).
WESMAEL, 1835, Nouv. Mem. Acad. Sci. Bruxelles 19: 86 (♀♂, Elassus).

Ephedrus japonicus ASHMEAD, 1906, Proc. U.S. Nat. Mus. 30: 187 (♀♂, Japan).
TAKAHASHI, 1925, Dept. Agric. Govt. Res. Inst. Formosa, Rept. 16: in
74 pp. Taiwan, host). GAHAN, 1926, Proc. U.S. Nat. Mus. 70:7 (Taiwan,
hosts – see TAKAHASHI, 1925). WATANABE, 1941, Ins. Mats. 15:136-9
(♀♂, Japan, Taiwan, hosts, notes). WATANABE, 1941, Ins. Mats. 15:170

(Japan, hosts). WATANABE, 1948, Ins. Mats. 21:1 (notes on the type).
NEW SYNONYMY

2. *Ephedrus plagiator* (NEES) var. *nigra* GAUTIER, BONNAMOUR & GAUMONT, 1929,
   Bull. Soc. Ent. Fr. 1929: 200-1 (France, host).

Diagnosis: This species is rather variable in body-length, wing venation, structure of tergite 1 and coloration, and is a polyphagous parasite of aphids. It differs from its congeners by the comparatively short antennal F_1 that is only slightly longer than F_2, by the long tergite 1 and by characters of wing venation.

Description – Female: Head transverse, rounded, about as wide as thorax, smooth, shiny, sparsely haired. Temple nearly equal to transverse eye-diameter. Gena as long as about $1/4$ of longitudinal eye-diameter. Eyes of medium size, oval, sparsely haired, a little convergent towards clypeus. Interocular line a little longer than transfacial line, shorter than facial line. Clypeus sparsely haired, with about 10 to 12 long hairs. Tentorio-ocular line equal to about $1/2$ of intertentorial line. Antennae 11-segmented, filiform, reaching apex of tergite 1. F_1 about 4 times as long as wide, F_2 a little shorter, (Fig. 149). Socket-ocular line equal to $1/2$ of socket-diameter.

Mesoscutum smooth, shiny, sparsely haired, notaulices wide, deep and rugose at the ascendent part, smooth afterwards but visible throughout. Propodeum (Fig. 113) areolated. Carinae forming central pentagonal areola of variable shape, which is sometimes divided irregularly by smaller carinae or rugosities (highly variable). Ventral carina and its bifurcation rather prominent. Discs of areolae smooth, slightly, irregularly rugose; along carinae with long scattered hairs. Wing (Fig. 156): Pterostigma about 4 times as long as wide. Radial abscissa 2 at least a little, but usually distinctly longer than interradial vein 1. Median abscissa 2 about twice as long as interradial vein 1, or a little shorter or longer. Radial cell 2 of quite variable shape.

Abdomen lanceolate. Tergite 1 (Fig. 146) slender, long, twice as long as wide at spiracles, more or less rugose, with two (more or less developed) lateral longitudinal carinae running out from base and sometimes with more or less distinct central longitudinal carina (the structure being rather variable), depressed apically and with two lateral impressions. Following tergites smooth, shiny, sparsely haired. Genitalia as figured (Fig. 34).

Coloration: Black. Base of F_1 and mouthparts brown. Legs brown to light brown, coxae brownish-black, femora and tibiae apically brownish-black, and tarsi obscured. Wing veins brown. Abdomen brownish-black, tergite 1 brown to brownish-black, tergites 2 and 3 with lighter spots.

Length of body about 1.7 to 3.5 mm.

Male: As described for female except for sexual differences.

General distribution: Probably all of the Palearctic and at least part of the Oriental region.

New material examined (681 ♂♀ specimens). Japan: Asakawa, April 25, 1961, ex. *Aphis gossypii* on *Callicarpa japonica*, 19 ♂♀ (E. I. SCHLINGER); Asakawa, April 27, 1961, ex. *Rhopalosiphoninus deutzifoliae* on *Deutzia crenata*, 6 ♀♀ (E. I. SCHLINGER); Asakawa, April 27, 1961, ex. *Parachaitophorus spiraeae* on *Spiraea nervosa*, 3 ♂♀ (E. I. SCHLINGER); Kagoshima, April 13, 1961, ex. *Macrosiphum* sp. on *Rosa* sp., 3 ♂♀ (E. I. SCHLINGER); Kagoshima, April 13, 1961, ex. *Aphis spiraecola* on *Spiraea thunbergi*, 98 ♂♀ (E. I. SCHLINGER); Kagoshima, April 13, 1961, ex. *Macrosiphum rosaeibarae* on *Rosa multiflora*, 4 ♂♀ (E. I. SCHLINGER); Kagoshima, April 12, 1961, ex. *Toxoptera odinae* on *Viburnum suspensum*, 4 ♂♀ (E. I. SCHLINGER); Tokyo, April 25, 1961, ex. *Aphis pomi* on *Chaenomelles sinensis*, 3 ♂♀ (E. I. SCHLINGER); Tokyo, May 13, 1961, ex. *Macrosiphum rosaeibarae* on *Rosa multiflora*, 3 ♀♀ (E. I. SCHLINGER); Tokyo, July 1912, ex. aphid on peach tree, 2 spec. (KUWATA, USNM); Fu-kuoka, April 18, 1961, ex. *Capitophorus* sp. on *Elaeagnus* sp., 8 ♂♀, (E. I. SCHLINGER) Fukuoka, April 2-17, 1961, ex. *Aphis spiraecola* on *Spiraea thun-bergi*, 145 ♂♀ (E. I. SCHLINGER); Osaka, April 21, 1961, ex. *Parachaitophorus spiraeae* on *Spiraea cantonensis*, 1 ♀ (E. I. SCHLINGER); Osaka, April 20, 1961, ex. *Aphis spiraecola* on *Spiraea thunbergi* and *S. cantonensis*, 45 ♂♀ (E. I. SCHLIN-GER); Gifu, Oct. 1902, 1 ♂ Paratype No. 7265 USNM (Y. NAWA). USSR (Pri-morye): Buchta Taungou, Sudzukhinsky zapovednik, Aug. 19, 1961, sandy habitat on sea shore, 6 spec. (TRIAPICYN); Stancia. Gorno-Taiozhnaja, July 9, 1961, edge of forest, 1 spec. (TRIAPICYN); Suputinsky zapovednik, Aug. 8, 1961, mixed forest, 3 spec. (KOVALEV & SHUVAKHINA); Vladivostok, Aug. 17, 1961, 2 spec. (TRIAPICYN); P. Tigrovyi, Suuansky rayon, Aug. 15, 1961, 1 spec. (PEREPELINA); Artjom, near Vladivostok, July 15, 1961, 3 spec. (TRIAPICYN). South Korea: Seoul, May 9, 1961, ex. *Amphorophora lonicericola* on *Lonicera* sp., 5 ♂♀ (E. I. SCHLINGER); Seoul, May 9, 1961, ex. *Aphis* sp. on *Prunus* sp., 1 ♂ (E. I. SCHLINGER). Hong Kong: Sha Tin, Feb. 28, 1961, ex. *Cavariella araliae* on *Aralia* sp., 140 ♂♀ (E. I. SCHLINGER). Taiwan: Taipci, March 5, 1961, ex. *Macrosiphum rosaeibarae* on *Rosa* sp., 92 ♂♀ (E. I. SCHLIN-GER); Taipei, March 6, 1961, ex. *Macrosiphum* sp. on *Rosa* sp., 3 ♂♀ (E. I.

SCHLINGER); Taipei, March 9, 1961, ex. *Megoura citricola* on *Munica* sp., 2 ♂ (E. I. SCHLINGER); Taipei, March 5, 1961, ex. *Toxoptera odinae*, 1 spec. (E. I. SCHLINGER); Taipei, March 20, 1961, ex. *Megoura citricola* on *Murraya paniculata*, 2 ♂♀ (E. I. SCHLINGER); Taipei, March 4, 1961, ex. *Aphis gossypii* on *Citrus* sp., 1 ♂ (E. I. SCHLINGER); Taipei, March 18, 1961, ex. *Aphis spiraecola* on *Spiraea* sp., 1 ♂ (E. I. SCHLINGER); Taichung, March 12, 1961, ex. *Myzus persicae* on *Nicotiana* sp., 1 ♂ (E. I. SCHLINGER); Wulai, March 21, 1961, ex. *Macrosiphum formosanum* on *Sonchus* sp., 7 ♂♀ (E. I. SCHLINGER); Wushe, March 12, 1961, ex. *Hyperomyzus lactucae* on *Lactuca sativa*, 1 ♂ (E. I. SCHLINGER); Yung Jean, March 11, 1961, ex. *Aphis spiraecola* on *Spiraea cantonensis*, 16 ♂♀ (E. I. SCHLINGER).

Type specimens:

"*Bracon plagiator* NEES" (*Aphidius parcicornis* NEES)

Type: Germany, Sickerhausen, "in Pruno ceraso horti" (lost).

Neotype ♀: Czechoslovakia, Hermanovce, Presovske hory, 15. V. 1959. bred from *Myzus cerasi* on *Prunus avium*, in an orchard (P. STARÝ), deposited in P. Starý coll.

Ephedrus japonicus ASHMEAD

Type: Cat. No. 7265 USNM, Gifu, Japan (Y. NAWA).

Habitat: In Europe it occurs mostly in forest-type habitats, however in Far East Asia it can be found in nearly all available habitats depending on the presence of aphid hosts.

Hosts: (Unrevised literature data).

Acyrthosiphon caraganae CHOL.: MEIER, 1927, USSR Eur. part., IVANOVA-KASAS, 1956.

Amphorophora magnoliae (ESSIG and KUWANA): WATANABE, 1941, on *Sambucus buergeriana*, Japan.

Amphicercus japonicus (HORI): WATANABE, 1941, on *Lonicera morrowii*, Japan.

Anuraphis mumei HORI: WATANABE, 1941, Japan.

Aphis laburni KALT.: WATANABE, 1941, on *Phaseolus vulgaris*, Japan.

Aphis evonymi FABR.: QUILIS M. P., 1934, on *Euonymus europaea*, Czechoslovakia.

Aphis pomi DEG.: REGNIER, 1923, LUZHETZKI, 1960, USSR – Uzbekistan.

Aphis medicaginis (KOCH): TELENGA, 1950, Eur. part of USSR.

Aphis rumicis L.: TAKAHASHI, 1925, Taiwan.

Brachysiphoniella graminis TAK.: TAKAHASHI, 1925, Taiwan.

Cavariella archangelicae (SCOP.) and *aegopodii* (SCOP.): PETERSEN, 1956, on *Archangelica officinalis*, Iceland. LINDROTH, 1931.

Dactynotus picridis (F.): TELENGA, 1950, Eur. part of USSR.

Dysaphis sorbi (KALT.): THUNEBERG, 1960, on *Sorbus aucuparia*, Finland.

Hyalopterus pruni FABR.: WIACKOWSKI & WIACKOWSKA, 1961, Poland. WATANABE, 1941, on *Prunus salicina*, Japan. TELENGA, 1950, Eur. part of USSR.

Macrosiphum rosae (L.): WATANABE, 1941, on *Rosa* sp., Japan.

Myzus cerasi F.: WIACKOWSKI & WIACKOWSKA, 1961, Poland.

Myzus persicae SULZ.: WATANABE, 1941, on *Raphanus sativus*, Japan.

Phorodon humuli (SCHRK.): TELENGA, 1950, Eur. part of USSR.

Prociphilus konoi HORI: WATANABE, 1941, on *Lonicera morrowii*, Japan.

Pyrethromyzus sanborni (GILL.): GAUTIER, BONNAMOUR & GAUMONT, 1929, France.[?]

Schizaphis longicauda HRL.: THUNEBERG, 1960, on *Phalaris arundinacea*, Finland.

Sitobium granarium (KIRBY): WATANABE, 1941, Japan.

Hosts: (Original and revised literature data).

Acyrthosiphon caraganae CHOL.: STARÝ, 1962, on *Caragana arborescens*, Czechoslovakia.

Acyrthosiphon spartii (KOCH): STARÝ, 1962, on *Sarothamnus scoparius*, Czechoslovakia.

★*Amphorophora lonicericola* TAKAH.: On *Lonicera* sp., South Korea.

Aphis farinosa (GMEL.): STARÝ, 1962, on *Salix* sp., Czechoslovakia.

Aphis craccae (L.): STARÝ, 1962, on *Vicia cracca*, Czechoslovakia.

★*Aphis gossypii* GLOV.: On *Citrus* sp., Taiwan. On *Callicarpa japonica*, Japan.

Aphis idaei (V. D. GOOT): STARÝ, 1962, on *Rubus idaeus*, Czechoslovakia.

Aphis fabae SCOP.: STARÝ, 1962, on *Euonymus europaea, Philadelphus coronarius, Borago officinalis, Impatiens roylei, Epipactis latifolia*, Czechoslovakia.

★*Aphis pomi* DEG.: On *Chaenomelles sinensis*, Japan.

★*Aphis spiraecola* PATCH: On *Spiraea thunbergi*, Japan. On *Spiraea* sp., Korea, Taiwan, and Japan.

Aphis sp.: STARÝ, 1962, on *Rhamnus cathartica*, Czechoslovakia.

Aphis sp.: STARÝ, 1962, on *Robinia pseudoacacia*, Czechoslovakia.

Asiphon tremulae (L.): STARÝ, 1962, on *Populus tremula*, Germany.

Brachycaudus cardui (L.): STARÝ, 1962, on *Prunus domestica, Prunus cerasifera, Prunus spinosa*, Czechoslovakia.

Capitophorus sp.: On *Elaeagnus* sp., Japan.

Cavariella araliae TAKAH.: On *Aralia* sp., S. China.

Ceruraphis eriophori (WALK.): STARÝ, 1962, on *Viburnum opulus*, Czechoslovakia.

Dysaphis devecta (WALK.): STARÝ, 1962, on *Malus silvestris*, Czechoslovakia.

Dysaphis sorbi (KALT.): STARÝ, 1962, on *Sorbus aucuparia*, Czechoslovakia.

Dysaphis spp.: STARÝ, 1962, on *Malus silvestris, Pyrus communis*, Czechoslovakia.

Hyperomyzus lactucae (L.): On *Lactuca sativa*, Taiwan. STARÝ, 1962, on *Ribes nigrum, Ribes grossularia*, Czechoslovakia.

Liosomaphis berberidis (KALT.): STARÝ, 1962, on *Berberis vulgaris*, Czechoslovakia.

Macrosiphum formosanum TAKAH.: On *Sonchus* sp., Taiwan.

Macrosiphum prenanthidis BOERNER: STARÝ, 1962, on *Prenanthes purpurea*, Czechoslovakia.

Macrosiphum rosae L.: STARÝ, 1962, on *Rosa* sp., Czechoslovakia. On *Rosa* sp., Poland.

Macrosiphum rosaeibarae MATS.: On *Rosa multiflora*, Japan. On *Rosa* sp., Taiwan.

Macrosiphum sp.: On *Rosa* sp., Taiwan, Japan.

Megoura citricola V. D. GOOT: On *Munica* sp., Taiwan. On *Murraya paniculata*, Taiwan.

Myzocallis coryli (GOETZE): STARÝ, 1962, on *Corylus avellana*, Czechoslovakia.

Myzus persicae SULZ.: On *Nicotiana* sp., Taiwan.

Myzus cerasi (F.): STARÝ, 1962, on *Prunus avium*, Czechoslovakia.

Parachaitophorus spiraeae TAKAH.: On *Spiraea cantonensis* and *S. nervosa*, Japan.

Phorodon humuli (SCHRK.): STARÝ, 1962, on *Prunus domestica*, Czechoslovakia.

Prociphilus fraxini (HTG.): STARÝ, 1962, on *Fraxinus excelsior*, Czechoslovakia.

Rhopalosiphoninus deutzifoliae SHIN.: On *Deutzia crenata*, Japan.

Rhopalosiphum infuscatum (KOCH): On *Malus* sp., Italy.

Rhopalosiphum padi (L.): STARÝ, 1962, on *Padus racemosa*, Czechoslovakia.

Schizoneura ulmi (L.): STARÝ, 1962, on *Ulmus laevis*, Czechoslovakia.

Sipha sp.: On *Festuca pratensis*, Eur. part of USSR.

Sitobium sp.: STARÝ, 1962, on *Secale cereale*, *Hordeum distichum*, *Dactylis glomerata*, Czechoslovakia.

Toxoptera odinae V.D. GOOT: On *Viburnum suspensum*, Japan, Taiwan.

Host-specificity: Widely polyphagous species.

Note: Parasitized aphid cocoons are usually dull to shiny black.

SUBGENUS *Lysephedrus* STARÝ

Ephedrus HALIDAY, subgenus *Lysephedrus* STARÝ, 1958, Acta Faun. Ent. Mus. Nat. Pragae 3: 65.

Type species: Aphidius (Ephedrus) validus HALIDAY as *Ephedrus validus* (HALIDAY), by original designation.

Differs from *Ephedrus* s. str. primarily by the coarsely rugose propodeum, structure of tergite 1 and by the densely pubescent ovipositor sheaths.

Description: Head transverse, no wider than thorax at tegulae. Antennae 11-segmented. Notaulices distinct, deeply rugose-punctate. Propodeum coarsely, irregularly and deeply rugose. Abdomen of female lanceolate. Tergite 1 coarsely rugose. Ovipositor sheaths very wide, rather rapidly and strongly narrowing from the proximal third towards apex; slightly curved upwards; entire surface densely pubescent, bluntly pointed at apex.

General distribution: Europe and South Korea.

Ephedrus (Lysephedrus) validus (HALIDAY)

Aphidius (Ephedrus) validus HALIDAY, 1833, Ent. Mag. 1: 485 (♀♂, England). THOMSON, C. G., 1895, Opusc. entomol. 25: 2332 (♀♂, Sweden). MARSHALL, 1896, in ANDRÉ, Spec. Hym. Eur. et d'Alg. 5: 542 (♀♂, England, host; *Ephedrus*). MARSHALL, 1899, Trans. Ent. Soc. London 1899:21 (♀♂, England, host). STELFOX, 1941, Ent. mon. Mag. 93:91 (Ireland, bion.). STARÝ, 1958, Acta Faun. Ent. Mus. Nat. Pragae 3:65-6 (♀♂, Czechoslovakia, Ireland). MACKAUER, 1961, Beitr. Ent. 11:99 (notes on the type).

Female: Head transverse, rounded, slightly narrowed behind eyes, smooth, shiny, densely haired. Temple nearly equal to transverse eye-diameter. Gena equal to $1/3$ of longitudinal eye-diameter. Eyes of medium size, oval, with sparse short hairs, slightly convergent toward the clypeus. Interocular line a little longer than transfacial line, shorter than facial line. Clypeus pubescent. Tentorio-ocular line, about $1/2$ of intertentorial line. Antennae 11-segmented, filiform, reaching apex of tergite 1, densely pubescent. F_1 five times as long as wide, two times longer than F_2. Socket-ocular line equal to socket-diameter.

Mesoscutum rugose in the ascendent part, densely pubescent. Notaulices deeply rugose especially in the ascendent part, distinct nearly up to praescutellar groove. Propodeum (Fig. 112) coarsely and deeply rugose, with long scattered hairs; with carinae forming an elevated areolation, but sometimes not distinct. Wing (Fig. 160): Pterostigma 4 to 5 times as long as wide. Radial abscissa 1 shorter than width of pterostigma. Radial abscissa 2 nearly equal to interradial vein 1. Median abscissa 2 about twice as long as interradial vein 1.

Abdomen lanceolate. Tergite 1 (Fig. 143) strongly rugose, similar to propodeum, with several more regular impressions before apex, with long scattered hairs. Spiracular tubercles little prominent, situated in the first third of the tergite. Following tergites smooth, shiny, more densely pubescent progressively towards apex. Genitalia as figured (Fig. 43), ovipositor sheaths densely pubescent.

Coloration: Blackish-brown; scape, pedicel, F_1 and sometimes basal part of F_2, mouthparts and sometimes lateral parts of tergite 1 yellowish to light brown. Legs brownish-yellow, tarsi apically obscured. Venation of wings light brown.

Length of body about 1.7 to 2.3 mm.

Male: As described for female except for sexual differences.

General distribution: Europe and South Korea.

New material examined (16 ♂♀ specimens). South Korea: Seoul, May 9, 1961, ex. Eriosomatine aphid on roots of *Apium* sp., 16 ♂♀ (E. I. SCHLINGER).

Holotype ♀: England, B. M. Type Hym 3. c. 70.

Habitat: This species was located in the Forestry Botanical Garden near Seoul on the roots of *Apium* sp. from 1″ to 2″ deep in loose soil. The mummies were black covered with considerable white flocculence.

Hosts: (Unrevised literature)

    Anuraphis farfarae KOCH: QUILIS, 1934, on *Malus silvestris*, Czechoslovakia.

    Aphis rumicis L.: WILKINSON, 1926, Cyprus.

62

Dysaphis pyri FONSC.: SAVARY, 1953, Switzerland.

Myzus cerasi (F.): MARSHALL, 1896, England.

Hosts: (Original and revised literature data)

*Eriosomatinae: On *Apium* sp. roots, South Korea.

GENUS LIPOLEXIS FÖRSTER

Lipolexis FÖRSTER 1862, Verh. Naturh. Ver. Preuss. Rheinl. 19:249

Type species: Lipolexis gracilis FÖRSTER

Gynocryptus QUILIS M. P. 1931, Eos 7:27-8.

Type species: Gynocryptus pieltaini QUILIS M. P.

Literary data: STARÝ, 1959, Acta Soc. Ent. Cechosl. 56:94 (revision of Eur. spp.). MACKAUER, 1962, Entomophaga, 7:43-4 (notes, key to world spp.).

This genus is easily distinguishable from its relatives by its wing venation and the shape of the female genitalia.

Description: Head transverse to subquadrate, wider than thorax. Occiput margined. Eyes large. Antennae filiform with variable number of segments (12-♀, 13-♂). Notaulices distinct at the ascendent part of mesoscutum. Propodeum areolated. Forewing: Pterostigma large, triangular; metacarp long. Radial vein long. Pterostigmal cell nearly complete. Cubital vein colourless, reaching wing-margin, otherwise venation effaced beyond basal cell towards the wing-apex. Hind wing without complete cells, only costal and subcostal vein developed. Abdomen of female lanceolate. Ovipositor sheaths curved downwards, more strongly sclerotized in upper part. Ovipositor curved downwards.

General distribution: Palearctic and Oriental regions.

Bionomics: Parasite of aphids with pupation occurring inside skin of parasitized aphid.

Key to the species of *Lipolexis* (♀♀)

1 Tergite 1 with strongly prominent central bifurcating carina (fig. 61). Coloration mostly brownish-black   *Lipolexis gracilis* FÖRSTER

  Tergite 1 without central carina. Coloration mostly yellow. 2

2 Ovipositor sheaths short and straight. Tentorio-ocular l. equal to $^1/_5$-$^1/_6$ of intertentorial line   *Lipolexis scutellaris* MACKAUER

  Ovipositor sheaths prolongated, somewhat dilated at apex. Tentorio-ocular line equal to $^1/_2$ of intertentorial line *Lipolexis oregmae* (GAHAN)

Lipolexis gracilis FÖRSTER

Lipolexis gracilis FÖRSTER 1862, Verh. Nat. Ver. Preuss. Rheinl. 19: 249 (♀♂, Germany). STARÝ, 1959, Acta Soc. Ent. Cechosl. 56: 94-6 (♀♂, Czechoslovakia, Germany, hosts). 1960, List d' Ident., No. 3, Entomophaga 5(4): 340 (France, host). – STARÝ, 1961, Acta Soc. Ent. Cechosl. 58: in 228-38 (bion., Czechoslovakia, hosts). STARÝ, 1962, Acta Soc. Ent. Cechosl. 59: 42-58 (ecol., Czechoslovakia, hosts).

Gynocryptus pieltaini QUILIS M. P. 1931, Eos 7: 27-8 (♀, Spain, host).

This species differs from the other two known species of the genus by the sculpturing of tergite 1.

Description – Female: Head (Figs. 56, 57) transverse to subquadrate, rounded, wider than thorax at tegulae, smooth, shiny. Occiput margined. Temple somewhat wider than $1/_2$ of transverse eye-diameter. Gena as wide as $1/_6$ of longitudinal eye-diameter. Eyes large, nearly hemispherical, prominent, sparsely haired, convergent towards clypeus. Interocular line $1/_2$ longer than transfacial line, somewhat shorter than facial line. Clypeus with 4 to 6 long hairs. Tentorio-ocular line, equal to $1/_5$ of intertentorial line. Antennae 12-segmented, filiform, reaching half of abdomen. F_1 equal to F_2, somewhat more than 3 times as long as wide. Socket-ocular line equal to socket-diameter.

Mesoscutum falling vertically to prothorax, smooth, shiny, with sparse long hairs along margins and effaced notaulices on the disc. Notaulices narrow, slightly crenulate, distinct at the ascendent part, effaced on the disc. Propodeum (Fig. 59) with wide more or less complete central areola. Wing (Fig. 183): Pterostigma more than 3 times as long as wide, as long as metacarp. Radial vein long, more than $1/_2$ longer than pterostigma.

Abdomen lanceolate. Tergite 1 (Fig. 61) slender, 3 times as long as wide at spiracles, with strongly developed central bifurcating carina, slightly rugose to nearly smooth. Spiracular tubercles barely visible. Following tergites smooth, shiny, sparsely haired. Genitalia as figured (Fig. 40). Ovipositor sheaths slender, long, slightly dilated at apex, more strongly sclerotized in the upper part.

Coloration: Head blackish-brown, base of antennae (scape, pedicel, F_1), clypeus and mouthparts yellow. Thorax brownish-black, prothorax and sometimes propodeum lighter colored. Wings lightly infuscated, venation brown. Legs yellow to light brown, upper part of femora, tibiae and tarsi slightly darkened. Abdomen brown, tergite 1 and base of tergite 2 usually yellow.

Length of body about 1.4 mm.

Male: Antennae 13-segmented. Coloration darker than in female, but otherwise as in female except for usual sexual differences.

General distribution: Europe (Czechoslovakia, Germany, France, Spain), USSR (Primorye), Japan, Hong Kong and Taiwan.

Material examined (34 ♂♀ specimens) USSR – Primorye: St. Okeanskaya, env. of Vladivostok, July 19, 1961, 1 spn., forest, (NIKOLSKAYA). Same locality, July 9, 1961, 2 spns., mixed forest, (KULOV). Same locality July 7, 1961, 1 spn., (KOZLOV). Akademgorodok, env. of Vladivostok, July 8, 1961, clearing in deciduous forest, 2 spns., (KOZLOV). Same locality, July 16, 1961, 1 spn., clearing in a forest, (KOZLOV). Same locality, July 27, 1961, 3 spns., (TRIAPICYN). Same locality July 16, 1961, 12 spns., swept on grasses, (TRIAPICYN). Suputinsky zapovednik, Aug. 3, 1961, mixed forest, 2 spns., (TRIAPICYN). Same locality Aug. 2, 1961, mixed forest, 1 spn., (SCHUVACHINA). Taiwan: Taipei, March 20, 1961, ex. *Toxoptera aurantii* on *Murraya paniculata*, 6 ♂♀ (E. I. SCHLINGER); Taipei, March 15, 1961, ex. *Aphis spiraecola* on *Spiraea* sp., 1 ♀ (E. I. SCHLINGER). Hong Kong: Kowloon, Feb. 25, 1961, ex. *Aphis gossypii* on *Hibiscus* sp., 1 ♀ (E. I. SCHLINGER). Japan: Fukuoka, April 6, 1961, ex. *Aphis spiraecola* on *Spiraea thunbergi*, 1 ♂, (E. I. SCHLINGER).

Type specimens:

Lipolexis gracilis FÖRSTER

♀, Boppard, 24/132, Först. cep. 1846. Deposited in coll. Förster in Zoological Museum of the Humboldt University at Berlin.

Gynocryptus pieltaini QUILIS

♀, Liria, Valencia (Hispania), QUILIS. Deposited in coll. Quilis, "Estación de entomología agrícola, Burjasot, Valencia, Spain".

Habitat: In Europe this species occurs in steppe (field) habitats from where it spreads into orchards and edges of woods. In Far East Asia this species was rather uncommon, but occurred most often in steppe areas.

Hosts: Unrevised literary data.

Aphididae sp.: QUILIS M. P. 1931, on *Pisum*, Spain.

Dysaphis plantaginea PASS.: List d' Identif. No. 3, Entomophaga, 1960, France.

Hosts: Original and revised literary data.

Anoecia sp.: On *Cornus sanguinea*, Czechoslovakia.

Aphididae sp.: STARÝ, 1959, on *Cirsium* sp., Czechoslovakia.

⋆*Aphis gossypii* GLOVER: On *Hibiscus* sp., S. China (Hong Kong).

Aphis (Cerosipha) bupleuri (BOERN.): On *Bupleurum falcatum*, Czechoslovakia.

Aphis (Pergandeida) craccae (L.): STARÝ, 1962, on *Vicia cracca*, Czechoslovakia.

Aphis (Pergandeida) craccivora (KOCH): STARÝ 1962, on *Medicago sativa*, *Onobrychis sativa*, Czechoslovakia.

Aphis fabae SCOP.: STARÝ, 1962, on *Beta vulgaris*, *Cirsium* sp., *Centaurea cyanus*, Czechoslovakia.

Aphis (Pergandeida)? *euphorbiae* (KALT.): On *Euphorbia cyparissias*, Czechoslovakia.

Aphis (Pergandeida) intybi KOCH: STARÝ, 1962, on *Cichorium intybus*, Czechoslovakia.

Aphis newtoni THEOB.: On *Iris variegata*, Czechoslovakia.

Aphis (Cerosipha) origani (PASS.): On *Origanum vulgare*, Czechoslovakia.

Aphis (Toxopterina) plantaginis (GOETZE): On *Plantago* sp., Czechoslovakia.

Aphis (Cerosipha) polygonata (NEVS.): On *Polygonum aviculare*, Czechoslovakia.

Aphis (Pergandeida) salviae (WALK.): STARÝ, 1962, on *Salvia nemorosa*, *S. pratensis*, Czechoslovakia.

⋆*Aphis spiraecola* PATCH: On *Spiraea thunbergi*, Japan, Taiwan.

Aphis (Toxopterina) taraxacicola (BOERNER): STARÝ, 1961, 1962, on *Taraxacum officinale*, Czechoslovakia.

Aphis (Toxopterina) sp.: STARÝ, 1961, on *Peucedanum alsaticum*, Czechoslovakia.

Aphis sp.: On *Rumex* sp., Czechoslovakia.

Brachycaudus cardui L.: STARÝ, 1959, 1962, on *Carduus* sp., Czechoslovakia.

Brachycaudus helichrysi KALT.: On *Melandrium album*, Czechoslovakia.

Brachycaudus mordwilkoi HRL.: STARÝ, 1961, 1962, on *Echium vulgare*, Czechoslovakia.

Brachycaudus sp.: STARÝ, 1962, on *Prunus domestica*, Czechoslovakia.

Brachycaudus sp.: On *Arctium lappa*, Czechoslovakia.

Brachycaudus sp.: On *Tanacetum* sp., Czechoslovakia.

Myzus cerasi (F.): STARÝ, 1962, on *Prunus cerasus*, Czechoslovakia.

Toxoptera aurantii (B. DE F.): On *Murraya paniculata*, Taiwan.

Host specificity: Widely oligophagous species, preferring distinctly *Aphis* species, sensu lato.

Lipolexis oregmae (GAHAN)

Diaeretus oregmae GAHAN 1932, Ann. ent. Soc. Amer. 25:736-7
(♀, Philippines, host); STARÝ, 1960, Ins. Matsum. 23(2): 109-11
(♀, Philippines, host; *Lipolexis*).

This species is similar to *L. scutellaris* MACKAUER, but is distinctly different by the shape of the ovipositor-sheaths.

Description – Female: Head transverse, smooth, shiny, sparsely haired, wider than thorax at tegulae. Occiput margined. Gena as wide as $^1/_4$ of longitudinal eye-diameter. Temple as wide as half of transverse eye-diameter. Eyes large, convex, with sparse and comparatively long hairs, somewhat convergent towards clypeus. Clypeus oval, convex, arcuate and margined frontally, with sparse long hairs, separated by shallow arcuate groove from face. Tentorio-ocular line equal to half of intertentorial line. Antennae 12-segmented, filiform, slender.

Thorax smooth, shiny. Mesoscutum gibbous, without covering pronotum when viewed laterally, with sparse long hairs. Notaulices distinct at the ascendent part, crenulate, effaced on the disc. Propodeum distinctly areolated (Fig. 60), central carina and its rami very prominent, nearly tuberculiform on the edges of the central trapezoidal large areola; separating one smaller upper and one large lower, slightly concave areola on each side; discs of areolae smooth, shiny, sparsely haired. Wing: Pterostigma triangular, about twice as long as wide.

Abdomen lanceolate. Tergite 1 (Fig. 63) very slender, about 2.5 times as long as wide at spiracles, smooth, shiny, slightly convex, sparsely haired. Spiracular tubercles situated about at midpoint of tergite, slightly prominent. Following tergites smooth, shiny, with long sparse hairs, more densely haired towards apex of abdomen. Ovipositor sheaths slender, downwards-curved, strongly narrowing to the apex and slightly dilated at apex. Ovipositor curved downwards (Fig. 46).

Coloration: Brownish-yellow. Head mostly yellow; frons, vertex, and occiput brown; mouthparts yellow. Antennae with scape, pedicel, F_1 and part of F_2 yellow, remainder brown. Mesoscutum and scutellum brown; the rest of thorax brownish-yellow, with more or less obscured tinges.

Legs yellow, praetarsi obscured. Tergite 1 brownish-yellow, remaining tergites brown.

Length of body about 2.6 mm.

Male: Unknown

General distribution: Philippine Islands.

Material examined: Philippines – Island of Panay, ex. *Oregma lanigera*, 2 ♀♀ paratypes No. 43433 USNM; 6186 red label, (A. W. LOPEZ, USNM).

Type: ♀, same data as examined paratypes (USNM).

Habitat: This species was not collected in areas of Far East Asia visited by SCHLINGER, although a species of a similar host aphid (*Oregma* sp.) was observed on banana near Taipei, Taiwan which was heavily parasitized by an aphidiid. The parasite could have been this species, but unfortunately all parasites had previously emerged.

Lipolexis scutellaris MACKAUER

Lipolexis scutellaris MACKAUER, 1962, Entomophaga 7(1): 43-4

(♀, Hong Kong, host).

This species is easily distinguished from its relative, *L. oregmae* (GAHAN), by the shape of ovipositor sheaths and sculpturing of tergite 1.

Description – Female: Head transverse to subquadrate, smooth, shiny, sparsely haired, wider than thorax at tegulae. Occiput margined. Temple somewhat wider than $1/_2$ of transverse eye-diameter. Gena as wide as $1/_6$ of longitudinal eye-diameter. Eyes large, nearly hemispherical, sparsely haired, strongly prominent, convergent towards clypeus. Interocular line $1/_2$ longer than transfacial line, somewhat shorter than facial line. Clypeus with 4 to 6 long hairs. Tentorio-ocular line equal to $1/_5$ to $1/_6$ of intertentorial line. Antennae 12-segmented, filiform, reaching to half of abdomen. F_1 equal to F_2, more than 3 times as long as wide. Socket-ocular l. equal to socket-diameter.

Mesoscutum falling vertically to prothorax, smooth, shiny, with sparse long hairs along margins and effaced notaulices on the disc. Notaulices distinct at the ascendent part, narrow, crenulate, effaced on the disc; fore margin being slightly prominent. Propodeum areolated (Fig. 58), with wide central areola, margins of which are strongly prominent. Wing (Fig. 182): Pterostigma triangular, twice as long as wide. Metacarp longer than pterostigma. Radial vein more than $1/_2$ as long as pterostigma.

Abdomen lanceolate. Tergite 1 (Fig. 64) slender, 2.5 to 3 times as long as wide at spiracles, slightly granulate, with slightly prominent lateral longitudinal carinae. Spiracular tubercles poorly visible. Following tergites smooth, shiny, with sparse long hairs. Genitalia as figured (Fig. 38).

Coloration: Head yellow, upper part of frons and occiput brown. Antennae brown, scape, pedicel and base of F_1 yellow. Thorax yellow, lobes with slight brown spots. Wings nearly hyaline, venation brown. Legs yellow, femora, tibiae and tarsi darkened. Tergite 1 yellow, the rest of abdomen brownish-yellow.

Length of body about 1.3 to 1.4 mm.

Male: Antenna 13-segmented. Coloration darker than in female, mostly light brown, with yellow face, mouthparts and prothorax.

General distribution: Hong Kong and Taiwan.

Material examined: (16♂♀ specimens): Hong Kong: Sha-Tin, May 7, 1955, ex. Aphid on pummelo (=*Citrus aurantium* ssp. *maxima*), 14 ♂♀, (S. K. CHENG). Taiwan: Taruko, March 18, 1961, ex. *Aphis malvoides*, 2 ♀ (E. I. SCHLINGER).

Holotype ♀: Tai Gu Shan, May 15, 1955, ex. aphid on pummelo, (S. K. CHENG), deposited in U.S. National Museum. The series of males cited above from Sha-Tin are certainly conspecific, but were unavailable for MACKAUER's study.

Habitat: This species was only encountered once in a citrus orchard in Taiwan.

GENUS LYSIPHLEBIA STARÝ AND SCHLINGER, NEW GENUS

This new genus is similar to *Lysiphlebus* FÖRSTER and *Lysaphidus* SMITH. It is related to *Lysiphlebus* FÖRSTER by the shape of the anterior prong of valvula 2, the shape of the ovipositor sheaths, characters on the head, and in wing-venation particularly the long metacarp. It differs from *Lysiphlebus* by (1) having the propodeum entirely areolated and strongly declivous in its lower part (sometimes in *Lysiphlebus* the propodeum has two divergent carinae in the lower part, but it is always without complete areola) and (2) by the shape and structure of tergite 1, which is coarsely rugose and which usually has a central longitudinal carina (in *Lysiphlebus* with more or less developed central tubercle only, without carina or coarse rugosities).

There is also a certain resemblance with *Lysaphidus* SMITH, such as the shape of tergite 1 and the central longitudinal carina, the sculpture of the propodeum which is entirely areolated (at least partly in *Lysaphidus*).

The characteristics of the head and the genitalia are, however, quite distinct. In *Lysaphidus* the general structure of the genitalia is related to *Aphidius* NEES, especially the remarkable, rather large anterior prong of valvula 2. Also, the metacarp is short in *Lysaphidus*.

Description: Head transverse, wider than thorax at tegulae. Eyes large. Tentorio-ocular line nearly equal to intertentorial line. Antennae filiform, with variable number of segments (13 to 16). Mesoscutum with notaulices distinct at the ascendent part, effaced on the disc. Propodeum entirely carinated, with medium sized (narrow) central areola. Fore wing: Pterostigma triangular. Metacarp long. Radial vein with two abscissae, 2nd interradial vein and rest of median vein distinct (under 2nd interradial vein). Radial cell 1 and median cell confluent, incomplete. Hind wing with complete basal cell. Tergite 1 slender or square, coarsely rugose, usually with longitudinal carina at the middle part. Ovipositor sheaths very slightly curved upwards, gradually narrowing and rounded at apex. Anterior prong of valvula 2 nearly straight, slightly dilated at apex.

Type species: *Lysiphlebus japonicus* ASHMEAD, by present designation.

General distribution: Japan, South Korea, Hong Kong and Taiwan.

Bionomics: Parasite of aphids with pupation occurring inside the parasitized aphid.

Key to the species of *Lysiphlebia*

Distance between spiracular tubercle and apex of tergite 1 about $1/2$ more than width at spiracles .... *L. japonica* (ASHMEAD)

Distance between spiracular tubercle and apex of tergite 1 a little shorter than width at spiracles .... *L. rugosa* STARÝ & SCHLINGER, n. sp.

Lysiphlebia japonica (ASHMEAD), new combination

Lysiphlebus japonicus ASHMEAD 1906, Proc. U.S. Nat. Mus. 30:190 (♀♂, Japan, host). GAHAN, 1926, Proc. U.S. Nat. Mus. 70:7 (Taiwan, Japan, host); WATANABE, 1957, Ins. Mats. 21:3 (notes on the type, Japan, host).

Description – Female: Head transverse, smooth, shiny, sparsely haired, strongly narrowed behind eyes, $1/5$ wider than thorax at tegulae. Occiput margined. Temple a little wider than half of transverse eye–diameter. Gena a little shorter than half of longitudinal eye–diameter. Eyes large, strongly prominent laterally (Fig. 106), widely oval, convergent towards clypeus, bald. Interocular line $1/5$ longer than transfacial line, of equal length to facial line. Clypeus oval with about 6 hairs, slightly convex, separated by shallow groove from face. Tentorio-ocular line a little shorter than inter-tentorial line. Antennae 13 to 14-segmented, slender, filiform, reaching to apex of tergite 1. F_1 $2^1/_2$ times as long as wide. $F_1 = F_2$. Antennal socket diameter twice as long as socket-ocular line.

Mesoscutum falling off roundly to prothorax, without covering it when viewed laterally, smooth, shiny, with sparse hairs along effaced notaulices on the disc. Notaulices deep, wide, crenulate, distinct at the ascendent part of mesoscutum only. Propodeum (Fig. 121) areolated, with prominent carinae that form narrow central areola, declivous behind longitudinal carinae. Wing (Fig. 179): Pterostigma $3^1/_2$ to $4^1/_2$ times as long as wide and as long as metacarp. Radial abscissa 1 and 2 of equal length, longer than width of pterostigma. Interradial vein 2 as long as $^1/_2$ of radial abscissa 1. Hairs on apical lateral margin longer than those on the surface.

Abdomen lanceolate. Tergite 1 (Fig. 122) $2^1/_2$-3 times as long as wide at spiracles, with feeble impression behind spiracular tubercles, somewhat convex at the hind part, dilating towards the apex, sparsely haired, some-what wider at apex than at spiracles, more or less coarsely rugose, with longitudinal central prominent carina that bifurcates at the second third. Spiracular tubercles more or less prominent. Distance between spiracular tubercle and apex about $^1/_2$ wider than width at spiracles. Remaining tergites smooth, shiny, sparsely haired. Genitalia as figured (Fig. 37).

Coloration rather variable: Head with frons, occiput, vertex, upper part of temples dark brown to brown. Face, lower part of temples, clypeus and mouthparts yellow. Antennae brown; scape, pedicel, flagellar segments 1 and 2 more or less yellow. Thorax brown, with yellow pattern; most of mesoscutum and scutellum dark brown, the rest of thorax yellowish-brown; or all the thorax yellow with brown mesoscutum. Abdomen brown; tergite 1, base of tergite 2 and suture between tergites 2 and 3 yellow.

Length of body about 1.7 to 2.1 mm.

Male: Antennae 16-segmented (17 according to ASHMEAD), long, filiform. Coloration similar to female. Hairs on apical lateral margin longer than those on the surface, otherwise as described for females.

General Distribution: Japan, Taiwan and South Korea.

Material examined (310♂♀ specimens). Japan: Tokyo, May 13, 1961, ex. *Brachycaudus helichrysi* on *Foeniculum vulgare*, 1 ♀ (E. I. SCHLINGER); Tokyo, June 21, 1917, ex. "orange aphid", 1 ♂ (K. YOSHIDA); Tokyo, Sept. 1902, ex. "black aphis on orange", 3 ♀ (I. KUWANA); Fukuoka, April 2 to 17, 1961, ex. *Aphis spiraecola* on *Spiraea thunbergi*, 38 ♂♀ (E. I. SCHLINGER); Osaka, April 20, 1961, ex. *Aphis spiraecola* on *Spiraea thunbergi*, 158 ♂♀ (E. I. SCHLINGER); Osaka, April 21, 1961, ex. *Aphis spiraecola* on *Spiraea cantonensis*, 3 ♂♀ (E. I. SCHLINGER); Kagoshima, April 10, 1961, ex. *Toxoptera odinae* on *Viburnum suspensum*, 1 ♂ (E. I. SCHLINGER); Angyo, 10 mi. N.W. of Tokyo, April 28, 1961, ex. *Toxoptera* sp. on *Rhus javanica*, 4 ♂♀ (E. I. SCHLINGER); Asakawa, April 27, 1961, ex. *Parachaitophorus spiraeae* on *Spiraea nervosa*, 3 ♂♀ (E. I. SCHLINGER). Taiwan: Taipei (Taihoku), Aug. 24, 1921, ex. "*Aphis sacchari*", 2 ♀ (*probably* coll. by R. TAKAHASHI). South Korea: Seoul, May 9, 1961, ex. *Tetraneura* sp. on *Ulmus* sp., 1 ♀ (E. I. SCHLINGER); Seoul, May 9, 1961, ex. *Aphis spiraecola* on *Spiraea* sp., 1 ♀ (E. I. SCHLINGER); Seoul, May 9, 1961, ex. *Aphis spiraecola* on *Crataegus pinnatigida*, 5 ♂♀ (E. I. SCHLINGER); Seoul, May 9, 1961, ex. *Aphis* sp. on *Sanguisorba officinalis*, 11 ♂♀ (E. I. SCHLINGER); Seoul, May 9, 1961, ex. *Aphis sambuci* on *Sambucus* sp., 1 ♂ (E. I. SCHLINGER); Chejudo Island, May 7, 1961, ex. *Aphis spiraecola* on *Citrus sinensis*, 39 ♂♀ (E. I. SCHLINGER).

Type specimen: Gifu, (Japan) Y. NAWA, Cat. No. 7271 USNM.

Habitat: This species was almost invariably encountered in forest or forest-transition areas and was usually associated with aphids living near to the ground. Although this species was often found parasitizing *Aphis spiraecola*, only the fundatrigenae of this species were attacked, indicating this host was a facticious one.

Hosts: (Unrevised literature data)

"orange aphis": GAHAN, 1926, Japan.

Aphis gossypii GLOV.: WATANABE, 1957, Japan.

Toxoptera aurantii (BOYER') : GAHAN, 1926, Taiwan.

Hosts: (Original and revised literature data).

Aphis sambuci L.: On *Sambucus*, S. Korea.

Aphis spiraecola PATCH: On *Citrus sinensis*, S. Korea, On *Spiraea thunbergi*, Japan. On *Spiraea cantonensis*, Japan. On *Spiraea* sp., S. Korea. On *Crataegus pinnatigida*, S. Korea.

Aphis sp.: On *Sanguisorba officinalis*, S. Korea.

Brachycaudus helichrysi (KALT.): On *Foeniculum vulgare*, Japan.

Longiunguis sacchari (ZEHNT.): Taiwan.

Parachaitophorus spiraeae TAKAH.: On *Spiraea nervosa*, Japan.

Tetraneura sp.: On *Ulmus* sp., S. Korea.

Toxoptera sp.: On *Rhus javanica*, Japan.

Toxoptera odinae v.d.g. On *Viburnum suspensum*, Japan.

"orange aphis": Japan.

Note: The specimens from Taiwan were kindly compared with the type series by c. f. w. muesebeck. The parasitized aphid cocoons are usually light brown.

Lysiphlebia rugosa starý and schlinger, new species

This species is closely related to *L. japonica* (ashm.), but differs from the latter by the shape of tergite 1, where the distance between the spiracular tubercle and apex of the tergite is shorter than (or nearly equal to) the width of the spiracles.

Description – Female: Head transverse, rounded, smooth, shiny, sparsely haired, wider than thorax at tegulae. Occiput slightly margined. Temple nearly as wide as transverse eye-diameter. Gena somewhat narrower than $1/3$ of longitudinal eye-diameter. Eyes large, widely oval, slightly convergent towards clypeus, bald. Interocular line a little longer than transfacial line (15: 13). Facial line a little longer than interocular line (16: 15). Clypeus with about 5 long hairs. Tentorio-ocular line a little shorter than intertentorial line (4:5). Antennae? -segmented (broken). Socket-ocular line $1/3$ shorter than socket-diameter.

Mesoscutum arcuately falling to prothorax, smooth, shiny, with sparse long hairs along effaced notaulices on the disc. Propodeum (Fig. 121) areolated; carinae prominent, some irregular, completing narrow central areola; spiracles on strongly prominent tubercles; sparsely haired. Wing (Fig. 179): Pterostigma $3^1/_2$ times as long as wide. Metacarp equal to length of pterostigma. Radial abscissa 1 equal to width of pterostigma.

Abdomen lanceolate. Tergite 1 (Fig. 116) less than twice as long as wide at spiracles; coarsely rugose to rugose-granulate, sometimes with central longitudinal carina between spiracular tubercles, which is, however, often hardly visible because of rugosities. Spiracular tubercles strongly prominent laterally and obliquely to above. Distance between spiracular tubercle and apex of tergite a little shorter than width at spiracles. Following tergites smooth, shiny, sparsely haired.

Coloration rather variable: Head brown, lower part of temples, face, clypeus and mouthparts yellow. Scape and pedicel yellow, F_1 brown. Thorax brown (part of mesoscutum, scutellum and metanotum, and part of mesopleurae), with variable yellow patterns. Wings hyaline, venation brown. Legs yellowish-brown, with brownish tinge on apices of femora, tibiae, and tarsi. Abdomen brown; tergite 1 and suture between tergites 2 and 3 yellow.

Length of body about 1.5 mm.

Male: Antennae 15-segmented, filiform. Hairs on apical lateral margin longer than those on the surface. Otherwise as described for female.

General distribution: Hong Kong and South Korea.

Holotype: ♀, Hong Kong, Taipo, February 28, 1961, ex. *Brachycaudus helichrysi* on unknown composite (E. I. SCHLINGER). The allotype and one ♂ paratype are topotypical. There are also 3 ♂ paratypes from Seoul, South Korea, May 9, 1961, ex. *Aphis* sp. on *Sanguisorba officinalis* (E. I. SCHLINGER).

Habitat: Both samples of the type material were collected in a transition-forest association.

GENUS LYSIPHLEBUS FÖRSTER

Lysiphlebus FÖRSTER 1862, Verh. Naturh. Ver. Preuss. Rheinl. 19:248, 250.

   Type species: *Aphidius (=Bracon) dissolutus* (NEES), (see STARÝ, 1961, notes).

Adialytus FÖRSTER 1862, Verh. Naturh. Ver. Preuss. Rheinl. 19:249, 250.

   Type species: *Adialytus tenuis* FÖRSTER.

Aphidaria PROVANCHER 1888, Addit. Corr. Faune Ent. Canada, p. 396.

   Type species: *Aphidaria basilaris* PROVANCHER.

Lysiphlebus FÖRSTER Subg. *Platycyphus* MACKAUER 1960, Beitr. Ent. 15: 590-1.

   Type species: *Lysiphlebus (Platycyphus) macrocornis* MACKAUER.

   Literary data: SMITH, 1944, Ohio State Univ. Contr. Zoo. Ent. 6:76-77 (rev. of Nearctic spp.). MACKAUER, 1960, Boll. Lab. Ent. Portici 18:230-6. (rev. of Nearctic spp.). MACKAUER, 1960, Beitr. Ent. 10: in 582-95 (rev. of European spp.). STARÝ, 1961, Bull. Ent. Pologne 31: in 97-103 (notes, catalogue of Eur. spp., *partim*). STARÝ, 1961, Acta Faun. Ent. Mus. Nat. Pragae 7: in 131-142 (notes and catalogue of European spp.).

This genus differs from other aphidiid genera primarily by features of the head, by wing-venation (with only the radial abscissa 1 and 2, and part of median vein or only radial vein developed), characters of the propodeum (usually entirely smooth or with two divergent carinae in the lower part) and the female-genitalia.

Description: Head transverse, wider than thorax at tegulae. Eyes of medium size. Tentorio-ocular line greater than $^1/_2$ of intertentorial line. Antennae filiform, with variable number of segments (12 to 17). Propodeum entirely smooth or with 2 short divergent carinae at lower part. Notaulices indistinct at base of mesoscutum. Fore wing: Pterostigma triangular, metacarp long. Radial vein with 2 abscissae or not divided, pterostigmal cell incomplete. Radial cell 1 and median cell confluent, incomplete or open. Interradial vein often colorless and distinct or effaced similar to rest of median vein. Abdomen lanceolate (female), or rounded (male). Tergite 1 triangular, smooth, sometimes with central tubercles. Ovipositor sheaths triangularly prolongated, slightly arcuate, slightly pointed to somewhat angular at apex. Anterior prong of valvula 2 comparatively short and slender, slightly dilated at apex, straight to slightly curved downwards. Ovipositor straight.

General distribution: Nearly cosmopolitan.

Bionomics: Parasite of aphids with pupation occurring inside skin of parasitized aphid.

Key to the species of *Lysiphlebus* (♀♀)

Fore wing with distinct interradial vein and part of median vein (Fig. 180)
    Tergite 1 widely triangular (Fig. 11)

L. sp., aff. *delhiensis* SUBBA RAO and SHARMA

Fore wing with effaced interradial and median vein (Fig. 181). Tergite 1 slenderly triangular (Fig. 123). (Parasite of *Chaitophorus* spp.)

L. salicaphis (FITCH)

Lysiphlebus sp., aff. *delhiensis* SUBBA RAO and SHARMA

Judging from the original description the following species is closely related to *L. delhiensis* SUBBA RAO and SHARMA (1960). *Delhiensis* is known as a parasite of *Longiunguis sacchari* (ZEHNT.) on sugarcane in India. The exact determination of our limited material has not been possible because of different criteria used in the original from those used in our descriptions.

Description – Female: Head transverse, rounded, slightly granulate to smooth, sparsely haired, wider than thorax at tegulae. Occiput slightly margined. Temple wider than transverse eye-diameter. Gena nearly as long as $^1/_2$ of longitudinal eye-diameter. Eyes small, oval, slightly convergent towards clypeus, unhaired. Interocular line somewhat longer than facial line as (15:13), $^1/_6$ shorter than facial line as (15:18). Clypeus with 8 long hairs. Tentorio-ocular line $^1/_3$ shorter than intertentorial line. Antennae broken. Socket-diameter $^1/_2$ longer than socket-ocular line.

Mesoscutum slightly granulate, with single hairs. Notaulices effaced on disc and poorly indicated at ascendent part. Propodeum (Fig. 120) smooth, with 3 to 4 hairs on each side. Wing (Fig. 180): Pterostigma prolongately triangular, 4 times as long as wide. Metacarp $^1/_2$ longer than pterostigma. Hairs on apical lateral margin longer than those on surface.

Abdomen lanceolate. Tergite 1 (Fig. 117) triangular, more than $^1/_2$ longer than wide at spiracles, dilating towards apex, smooth, shiny, slightly convex, with poorly distinct central tubercle near base. Spiracular tubercles hardly prominent. Width at spiracles equal to distance between spiracles and apex. Width at apex equal to length. Following tergites smooth, shiny, sparsely haired. Genitalia as figured (Fig. 41).

Coloration: Head dark brown, neighborhood of mandibles and mouth-parts yellow. Antennae brown, scape and pedicel yellowish-brown. Thorax brown, apex of propodeum yellow. Wings hyaline, venation brown. Legs brown, trochanters and bases of tibiae yellow. Tergite 1 and basal part of tergite 2 yellow, rest of abdomen brown.

Length of body about 1.5 mm.

Male: Antennae broken. Hairs on apical lateral margin of fore wings longer than those on the surface. Otherwise like female except for usual sexual differences.

General distribution: South Korea.

Material examined: South Korea: Seoul, May 9, 1961, ex. *Aphis* sp. on *Sanguisorba officinalis*, 2 ♀, 1 ♂. (E. I. SCHLINGER).

Habitat: This species was only encountered one time during the Far East Asian trip and that was in a forest habitat on the Forest Experiment Station grounds.

Lysiphlebus salicaphis (FITCH)

Trioxys salicaphis FITCH, 1855 (1854). New York St. Agr. Soc. Trans. 14: 841.

Diaeretus salicaphis (FITCH): SMITH, 1944, Aphidiinae of N. A., p. 102; MUESEBECK, et al., 1951. USDA Agr. Monog. 2, p. 97.

Aphidius (Diaeretus) laticephalus TELENGA 1953, Trudy Inst. Zool. i parazitol. AN Uzb. SSR 1: 172-3 (♀, U.S.S.R.-Uzbekistan, host). LUZHETZKI, 1959, Tez. Dokl. 4-ogo Vses. Ent. Obstsh. Leningrad, p. 82 (USSR – Uzbekistan). LUZHETZKI, 1960, Par. tlej Uzbekistana, p. 35-6 (♀♂. USSR – Uzbekistan, host; *Diaeretus*). STARÝ, 1961, Bull. ent. Pologne 31: in 97-103 (notes, *Lysiphlebus*). NARZYKULOV & ATAEVA, 1961, Trudy Inst. Zool. Parazitol. Tadj. SSR 20: 189-90 (Tajikistan, hosts). (New Synonymy).

Description – Female. Head widely transverse, shiny, feebly granulate, wider than thorax at tegulae. Occiput margined. Temple about $^1/_4$ narrower than transverse eye-diameter. Gena as long as about $^1/_4$ of longitudinal eye-diameter. Eyes large, widely oval, convex, slightly convergent towards clypeus. Clypeus oval, convex, smooth, shiny, with about 6 hairs, separated by shallow arcuate groove from face. Tentorio-ocular line $^1/_4$ shorter than intertentorial line. Antennae 12 to 13-segmented, (14 according to the original description), as long as head, thorax and tergite 1 together. F_1 and F_2 of equal length, 2.5 times as long as wide.

Mesoscutum without covering prothorax when viewed laterally, smooth and shiny, along margins and effaced notaulices on the disc with sparse long hairs. Notaulices distinct, deeply crenulate at fore part, effaced on disc. Propodeum smooth, with few hairs on upper part. Wing (Fig. 181): Pterostigma triangular, about 3 times as long as wide. Metacarp somewhat longer than pterostigma. Radial vein as long as 2.5 times pterostigma width.

Abdomen lanceolate. Tergite 1 (Fig. 123) about 2.5 times as long as wide at spiracles, strongly narrowed before spiracular tubercles, with feeble lateral impressions behind them, and slightly dilating towards apex; comparatively flat, smooth, shiny, with keel-form protuberance at basal third; sparsely haired at hind part. Spiracular tubercles slightly prominent, situated somewhat before half of tergite. Genitalia as figured (Fig. 39).

Coloration: Head brownish-black; face more or less, clypeus and mouthparts yellowish-brown. Antennae brownish-black, base of F_1 yellow. Thorax brownish-black, prothorax sometimes yellow. Wing venation brown. Legs yellowish-brown, tarsi and upper part of hind femora somewhat darkened. Abdomen: Tergite 1 and base of tergite 2 yellow, rest of abdomen brown.

Length of body 1.5 to 1.7 mm.

Male: Antennae 15-segmented, nearly as long as the body. Tergite 1 (Fig. 124): central keel seems to be less separated and tergite less narrowed than in female. Otherwise like female except for usual sexual differences.

General distribution: Europe (Czechoslovakia, Hungary, Bulgaria), Central Asia (USSR – Uzbekistan, Tajikistan) and South Korea.

Material examined: South Korea: Seoul, May, 1961, ex. *Chaitophorus salicicolus* on *Salix* sp., 2 ♀♀. (E. I. SCHLINGER).

Type specimen: USSR, Uzbekistan, Ak-Tepe, Surkhan-darya, June 15, 1927, ♀, bred from aphids on *Salix* sp., coll. Ber. Probably deposited in the collection of Prof. TELENGA at Kiev. *(L. laticephalus* TEL.)

Habitat: This species was only encountered on one occasion in a forest situation at the Forest Experiment Station near Seoul.

Hosts: (Unrevised literary data)

Aphidae sp.: TELENGA, 1953, on *Salix*, USSR – Uzbekistan.

Hosts: (Original and revised literary data)

Chaitophorus albus MORDV.: NARZYKULOV & ATAEVA, 1961, on *Populus alba*, USSR – Tajikistan.

Chaitophorus leucomelas KOCH: NARZYKULOV & ATAEVA, 1961, USSR – Tajikistan.

*Chaitophorus salicicolus MATS.: On *Salix* sp., S. Korea.

Chaitophorus salicivorus WALK.: NARZYKULOV & ATAEVA, 1961, on *Salix caprea*, USSR – Tajikistan.

Chaitophorus sp.; On *Populus* sp., Bulgaria.

Neothomasia populicola BAK.: LUZHETZKI, 1960, on *Populus nigra*, USSR – Uzbekistan.

Host specificity: This parasite appears to be oligophagous but apparently prefers mostly hosts of the genus *Chaitophorus*.

GENUS MONOCTONUS HALIDAY

Aphidius NEES subg. *Monoctonus* HALIDAY 1833, Ent. Mag. 1:261, 487.

Type species: Aphidius (Monoctonus) caricis HALIDAY.

Subgenera

1. *Falciconus* MACKAUER

   Monoctonus HALIDAY subg. *Falciconus* MACKAUER 1959, Senck. Biol., Frankfurt M., 40:180

   Type species: Aphidius pseudoplatani MARSHALL

2. *Monoctonus* s. str.

   Monoctonus HALIDAY subg. *Monoctonus* s. str., STARÝ 1959, Acta Soc. Ent. Cechosl. 56: 241-2

   Type species: Aphidius (Monoctonus) caricis HALIDAY

3. *Paramonoctonus* STARÝ

Monoctonus HALIDAY subg. *Paramonoctonus* STARÝ 1959, Acta Soc. Ent. Cechosl. 56:238-9.

Type species: Monoctonus (Paramonoctonus) angustivalvus STARÝ

Literary data: SMITH, C. F., 1944, Ohio State Univ. Contr. Zoo. Ent. 6:33 (revision of Nearctic spp.). STARÝ, 1959, Acta Soc. Ent. Cechosl. 56:237-8 (revision of European spp.).

Description: Head subcubical. Eyes large, oval. Antennae 13 to 18-segmented, filiform. Notaulices distinct at fore part of mesoscutum. Propodeum with distinct central pentagonal areola or only with two divergent carinae in lower part. Fore wing: Pterostigma triangular. Radial and median cells confluent and more or less distinctly separated by intermedian + median vein and interradial vein on outer sides, sometimes only radial vein developed beyond basal cell. Radial vein always developed, first abscissa, if distinguishable, perpendicular or oblique to pterostigma. Hind wing with complete basal cell. Abdomen lanceolate or rounded. Ovipositor curved downwards. Ovipositor sheaths curved downwards, ploughshare-shaped or only slightly dilated in center and narrowed towards the apex.

General distribution: Holarctic and parts of Oriental region.

Bionomics: Parasite of aphids with pupation occurring inside skin of parasitized aphid.

Note: Both species known from Far East Asia belong to the subgenus *Monoctonus.*

Key to the species of *Monoctonus (Monoctonus)* (♀♀)

Antennae 13-segmented. Ovipositor sheaths ploughshare-shaped. Tergite 1, $1/_2$ longer than width at spiracles. Coloration of body mostly brown
                                      Monoctonus (M.) similis STARÝ and SCHLINGER, n. sp.

Antennae 16-segmented. Ovipositor sheaths slender. Tergite 1, 3.5 times as long as wide at spiracles. Coloration of body mostly yellow
                                  Monoctonus (M.) woodwardiae STARÝ and SCHLINGER, n. sp.

Monoctonus (Monoctonus) similis STARÝ and SCHLINGER, new species

This new species is related to *Monoctonus cerasi* (MARSH.) and *M. caricis* (HAL.) by the 13-segmented antennae. It differs from *M. cerasi* by the short notaulices, the more densely pubescent mesoscutum, the characters on tergite 1, the female genitalia and by coloration. It is distinguishable from *M. caricis* by the short notaulices, the more densely pubescent mesoscutum, the lanceolate abdomen of female, the characters of the female genitalia and by coloration.

Description – Female: Head transverse, rounded, smooth, shiny, sparsely haired, wider than thorax at tegulae. Occiput slightly margined. Temple $1/5$ narrower than transverse eye-diameter. Gena very narrow, as wide as $1/10$ of longitudinal eye-diameter. Eyes large, oval (width to length as 2:3), sparsely and shortly haired, convergent towards clypeus. Interocular line $1/2$ longer than transfacial line, about $1/4$ shorter than facial line. Facial line twice as long as transfacial line. Clypeus convex, with 9 hairs. Tentorio-ocular line as long as $1/6$ of intertentorial line. Antennae 13-segmented, fili-form, densely haired, reaching apex of tergite 1. F_1 nearly 4 times as long as wide. F_2 somewhat more than 3 times as long as wide. Socket-ocular line equal to socket-diameter.

Mesoscutum falling comparatively vertically to prothorax, without covering it when viewed laterally, smooth, shiny, with sparse long hairs in ascendent part of lobes, along margins and along effaced notaulices. Notaulices feebly crenulate, distinct at ascendent part of mesoscutum only and effaced on disc. Propodeum (Fig. 126) areolated; discs of areolae shiny, sparsely haired, with variably developed irregular rugosities to carinae. Wing (Fig. 162): Pterostigma long and triangular, nearly 3 times as long as wide. Metacarp as long as $1/3$ of pterostigma-length. Radial abscissa 1 and 2 distinct, colored; interradial vein and median vein (completing confluent radial and median cells) colorless but distinct.

Abdomen lanceolate. Tergite 1 (Fig. 128) $1/2$ longer than width at spiracles; rugose, with prominent bifurcating central carina, sparsely haired. Spiracular tubercles slightly prominent laterally, situated before middle of tergite. Width at apex less than at spiracles. Following tergites smooth, shiny, with sparse long hairs. Genitalia as figured (Fig. 31).

Coloration: Head brownish-black; clypeus, lower part of genae and mouthparts yellowish-brown; antennae dark brown, scape, pedicel and the main part of F_1 yellow. Thorax brown; prothorax, upper part of mesopleurae, metapleurae and propodeum lighter brown. Wings hyaline, venation mostly light brown. Legs yellowish-brown; upper base of femora, apices of tibiae and tarsi darkened. Tergite 1 and base of tergite 2 yellow. Following tergites brown. Ovipositor sheaths brownish-yellow.

Length of body about 3.5 mm.

Male: Antennae 15 to 16-segmented. Coloration about as in female but usually somewhat darker, otherwise as described for female except for usual sexual differences.

General distribution: Japan.

Holotype ♀: Japan, Noboribetsu, May 20, 1961, ex. *Myzus* sp. on unknown, young composite plant (E. I. SCHLINGER). Allotype and 2 ♀ paratypes are topotypical.

Habitat: The type locality consisted of a cleared hillside area in a rather dense larch and pine forest situation.

Monoctonus (Monoctonus) woodwardiae STARÝ and SCHLINGER, new species

This new species is somewhat related to the European *M. (M.) cerasi* (MARSH.) in regards to the shape of ovipositor sheaths and host-complex, but it is easily distinguishable from the latter by such features as the number of antennal segments, the wing venation and the structure of tergite 1.

Description – Female: Head subcubical, rounded, wider than thorax at tegulae, smooth, shiny, sparsely haired. Occiput margined. Temple a little narrower than transverse eye-diameter. Gena as wide as about $^1/_5$ to $^1/_7$ of transverse eye-diameter. Eyes large, oval, sparsely shortly haired, convergent towards clypeus. Interocular line nearly twice as long as transfacial line, nearly equal to facial line. Clypeus oval, with 6 long hairs. Tentorio-ocular line as $^1/_4$ of intertentorial line. Antennae 16-segmented, filiform, reaching half of abdomen. F_1 and F_2 equal, nearly 4 times as long as wide. Socket-ocular line about $^1/_2$ of socket-diameter.

Mesoscutum falling almost vertically to prothorax, without covering it when viewed laterally; smooth, shiny, with long hairs along lateral margins and on effaced notaulices on disc. Notaulices distinct at ascendent part, wide, transversely crenulate, effaced on disc. Propodeum (Fig. 114) areolated, with central, comparatively wide areola. Wing: Pterostigma triangular, about 7 times as long as wide. Metacarp short, as long as $^1/_3$ to $^1/_4$ of pterostigma. Radial abscissa 1 straight, nearly perpendicular to pterostigma, a little shorter than width of pterostigma. Radial abscissa 2 about 3 times as long as abscissa 1. Interradial vein 2 distinct.

Abdomen lanceolate. Tergite 1 (Fig. 125) slender, 3.5 times as long as wide at spiracles, with irregular carinae, granulo-rugose and sparsely haired. Spiracular tubercles hardly prominent. Following tergites smooth, shiny, with sparse long hairs. Genitalia as figured (Fig. 30). Ovipositor sheaths' comparatively slender.

Coloration: Head brownish-black, face brownish-yellow, lower part of temples, genae and mouthparts yellow. Scape, pedicel, base of F_1 yellow, the rest of flagellum brown. Thorax orangish-yellow. Wings hyaline, venation light brown. Legs yellow, praetarsi darkened. Tergite 1, basal spot of tergite 2 and apex of abdomen yellow, the rest of abdomen brown. Ovipositor sheaths light yellow.

Length of body about 2.1 mm.

Male: Unknown.

General distribution: Taiwan

Holotype ♀: Taiwan, Wulai, March 21, 1961, ex. *Myzus woodwardiae* on *Woodwardia* sp., (E. I. SCHLINGER). There are also 2 ♀♀ paratopotypes.

Habitat: The type locality consisted of a very humid roadside area where *Woodwardia* ferns were growing out from beneath and between rocks. The surrounding area was shrub-forest with mixed agriculture.

GENUS PARALIPSIS FÖRSTER

Paralipsis FÖRSTER 1862, Verh. Nat. Ver. Preuss. Rheinl. 19:248.

Type species: Aphidius enervis NEES.

Myrmecobosca MANEVAL 1940, Bull. Soc. Linn. Lyon 9:9.

Type species: Myrmecobosca mandibularis MANEVAL.

Literary data: STARÝ, 1958, Acta Faun. Ent. Mus. Nat. Pragae 3:85-9. (revision of European species, notes).

This genus is rather remarkable by its robust appearance, and is easily separated from other genera by its wing venation.

Description: (Fig. 152) Head nearly square as seen from above, vertex flat, face vertical. Eyes rather small. Antennae 15 to 18-segmented, filiform, very strong. Notaulices distinct at fore part. Propodeum smooth. Wing venation very strong, well developed, with costal, subcostal, basal, cubital and anal veins present. Radial vein visible only as point on lower margin of pterostigma. Pterostigma large, triangular, strongly sclerotized. Hind wing with complete basal cell. Legs strong. Abdomen of female widely oval. Ovipositors heaths straight, short, relatively wide. Ovipositor straight.

General distribution: Europe and Japan.

Bionomics: Parasite of certain root feeding aphids which are protected by ants. Pupation occurs inside skin of parasitized aphid.

Paralipsis eikoae (YASUMATSU)

Myrmecobosca eikoae YASUMATSU 1951, Rev. franç. d'Ent. 18:172-4.

(♀, Japan, bion.). STARÝ, 1958, Acta Faun. Ent. Mus. Nat. Pragae 3:89 (notes on *Paralipsis*). YASUMATSU, 1960, Kontyu 28:57 (add. notes).

This species differs from the European *P. enervis* (NEES) by the following criteria (partly after YASUMATSU, 1960): Head distinctly wider than long as seen from above. Distance between posterior ocelli 4 times as long as diameter of ocellus. Cubital cell 2 not at all defined. Tergite 1 remarkably longer than wide, the sides distinctly divergent posteriorly.

Description – Female (partly after YASUMATSU, 1960): Head (Fig. 155) as high as long when seen laterally, distinctly wider than high when seen frontally. Vertex flattened horizontally in right angle to frons which drops vertically to mouth opening. Vertex longer than frons, with central proecellar impression. Occiput slightly rounded. Genae long, rounded at the hind part. Eyes pubescent, oval, much longer than genae. Clypeus cushion-shaped, prominent, nearly twice as wide as long, very feebly rounded frontally, separated by deep groove on each side. Ocelli rather small, situated at the hind part of vertex; distance between hind ocelli a little longer than distance between fore and hind ocellus. Mandibles strong, prominent frontally, with long hairs at the fore part, bidentate at apex. Maxillary palpi 2-segmented, the last segment ovally prolongated, with long hairs. Antennae 16-segmented, reaching nearly to apex of tergite 2, their insertion surrounded by a convexity; scape arcuated, second segment short, as long as wide; funiculus nearly subfiliform, somewhat more slender at base than at apex; 3rd, 4th and 16th segments as long as wide, 5th segment less than twice as long as wide, remaining segments nearly equal; segments prolongately striated up to 7th segment.

Thorax seems gibbous, its highest point much higher than the head-level. Prothorax invisible as seen from above. Mesoscutum convex, wider than long, deeply lowered cephalad, pubescent. Scutellum with transverse wide and deep groove, carinated laterally; triangular, flat, rounded at sides. Propodeum transverse, widely rounded at hind angles. Pleurae without other sculptures except feeble and sparse punctuation. Fore·wing (Fig. 153) without cubital cell 2. Hind wing with 3 hamuli. Wings pubescent.

Tergite 1 (Fig. 154) somewhat longer than propodeum, somewhat longer than wide at apex, dilating gradually to the apex, spiracular tubercles almost tipped.

Coloration: Head, thorax and abdomen brown, shiny, but not entirely smooth. Antennae, frons, mandibles, clypeus, legs, pleurae, propodeum, tergites 1 and 2 light yellowish-rufous. Wings infumated. Nervation and pterostigma dark brown.

Length of head and thorax nearly 1 mm.

Male: Unknown.

General distribution: Japan.

Type specimen: ♀, Japan, Kyushu, Mt. Hiko, Japan (in collection of Kyushu University).

Note: This species was redescribed and differentiated from *P. enervis* NEES more exactly by YASUMATSU in 1960, since STARÝ had thrown doubt on its validity in 1958. For this reason we accept this as a valid species although we have not examined any specimens personally.

GENUS PAUESIA QUILIS

Pauesia QUILIS M. P. 1931, Eos, 7:67-9.

Type species: Pauesia albuferensis QUILIS

Aphidius NEES subg. *Paraphidius* STARÝ 1958, Acta Faun. Ent. Mus. Nat. Pragae 3:56, 91.

Type species: Aphidius californicus ASHMEAD

Literary data: STARÝ, 1960, Acta Faun. Ent. Mus. Nat. Pragae 6:5-44 (rev. of Eur. spp. and cat. of world spp.).

This genus differs from its relatives by wing-venation, sculpture of propodeum and host complex (parasites of various species of Lachnidae exclusively).

Description : Head transverse, wider than thorax at tegulae. Face broad. Eyes large, often hemispherical, strongly prominent laterally. Width of head subequal to twice that of transfacial line. Antennae filiform, with various numbers of segments (16 to 31). Mesoscutum with more or less distinct notaulices. Fore wing: Pterostigma triangular, strongly sclerotized. Metacarp distinctly developed, always longer than width of pterostigma. Radial vein distinct, with two abscissae. Pterostigmal cell incomplete. Radial and median cells confluent and completed by fused intermedian and median veins on the lower side and by interradial vein on external side. Hind wing with complete basal cell. Propodeum: Rami of central carina more or less developed, sometimes more or less effaced on longitudinal part, completing more or less concave, wide, central areola. Abdomen of female lanceolate. Oviposi-

84

tor sheaths of various shapes, more or less curved upwards to nearly straight.

General distribution: Holarctic, Oriental and Ethiopian regions.

Bionomics: Parasite of aphids (Lachnidae). Pupation occurs inside skin of parasitized aphid.

Key to the species of *Pauesia* (♀♀)*

1 Ovipositor sheaths slender, strongly narrowing towards apex (Fig. 45)
  Antennae 16 to 17-segmented   *Pauesia unilachni* (GAHAN)
  Ovipositor sheaths wide, stout, only slightly narrowed towards apex
  (Fig. 48) 2
2 Apical part of flagellum white   *Pauesia infulata* (HALIDAY)
  Flagellum entirely black 3
3 Tergite 1 narrowed behind spiracular tubercles, twice as wide at apex
  as at spiracles (Fig. 81)   *Pauesia pini* (HALIDAY)
  Tergite 1 gradually dilated towards apex (Fig. 83), half as wide at apex
  as at spiracles   *Pauesia tropicalis* STARÝ and SCHLINGER, n. sp.

Pauesia infulata (HALIDAY)

Aphidius (Aphidius) infulatus HALIDAY 1834, Ent. Mag. 2:96 (♀♂, England). MARSHALL, 1896, in ANDRÉ, Spec. Hym. Eur. et d'Alg. 5:564-5 (♀♂, England), MARSHALL,1899, Trans. ent. Soc. London 1899:39 (♀♂, England). FAHRINGER, 1937, Festschr. 60. Geb. E. Strand. Riga 3:241, 243 (descr., North and Central Europe, host). TELENGA, 1948, Trudy Inst. zool. AN Ukr. SSR 1:155 (USSR – Ukraine, in forest habitats). STARÝ, 1960, Acta Faun. ent. Mus. Nat. Pragae 6:14-16 (♀♂, Czechoslovakia, USSR – Eur. part; hosts; as *Paraphidius)*. MACKAUER, 1961, Beitr. Ent. 11:107 (notes on the type).

Paraphidius albiflagellaris STARÝ 1960, Acta Faun. Ent. Mus. Nat. Pragae 6: 10-11 (♀, Czechoslovakia). STARÝ, 1962, Acta Soc. Ent. Cechosl. 59: in 42-58 (ecol., hosts in Czechoslovakia).

This species differs from its relatives chiefly by the coloration of the flagellum.

Description – Female: Head (Fig. 51) transverse, smooth, shiny, wider than thorax at tegulae, sparsely haired, strongly narrowed behind eyes. Occiput distinctly margined. Temple as wide as ¹/₃ of transverse eye-

* Note that only species studied by the authors are included in the key, but other species are described later on.

diameter. Gena as about $^1/_5$ of longitudinal eye-diameter. Eyes large, hemispherical, strongly prominent laterally, sparsely haired, slightly convergent towards clypeus. Interocular line $^1/_3$ longer than transfacial line, shorter than facial line. Clypeus slightly granulate, with sparse long hairs. Tentorio-ocular line equal to half of intertentorial line. Antennae 20 to 21-segmented, filiform, reaching about the apex of tergite 1. F_1 a little longer than F_2. Socket-ocular line equal to half of socket-diameter.

Mesoscutum raised above prothorax without covering it when viewed laterally, shiny, granulate, sparsely haired. Notaulices distinct at the ascendent part, crenulate, effaced on the disc. Propodeum (Fig. 55): Rami of the central carina prominent and distinct especially in the transverse part, completing central pentagonal wide areola and separating one large upper and one lower areola on each side. Discs of areolae coarsely rugose, shiny, only central areola somewhat smoother; with sparse long hairs. Wing (Fig. 165): Pterostigma triangular, strongly sclerotized, longer than metacarp. Radial abscissa 1 somewhat longer than width of pterostigma.

Abdomen lanceolate. Tergite 1 (Fig. 84) about 3 times as long as wide at spiracles, rugose-granulate, with sparse long hairs, with lateral impressions beyond spiracular tubercles and slightly dilated to apex. Spiracular tubercles slightly prominent, situated somewhat before middle of tergite. Following tergites smooth, shiny, sparsely haired. Genitalia as figured (Fig. 48).

Coloration: Head yellowish-rufous, upper part (frons partially, occiput and upper part of temples) black; mandibles with dark apices. Antennae: scape and sometimes pedicel yellowish-rufous, flagellar segments black, apical part (beyond 12th segment, but variable) white, the last segment usually brown. Thorax brownish-black, prothorax yellowish-rufous. Wings hyaline, venation brown, fore half of costal and subcostal veins white. Tegulae yellow. Fore legs yellowish-rufous, tarsi obscured. Middle and hind legs yellowish-rufous, coxae, upper part of femora, apices of tibiae and tarsi obscured. Abdomen: Tergite 1 blackish-brown, tergite 2 brown; remaining tergites yellowish-rufous to yellow with brown apices. Ovipositor sheaths dark brown.

Length of body about 2.5 mm.

Male: (Acc. to HALIDAY) black; antennae black. Wings hyaline, tegulae dull stramineous, venation fuscous, fore legs stramineous, more dull on outer side; middle and hind legs fuscous with almost whole of trochanters, both ends of tibiae, and base of tarsi stramineous; all coxae black; abdomen piceous with luteous patch in middle above; 1st segment scarcely dilated at extremity.

General distribution: Europe and Japan.

Material examined: Japan: Nopporo, May 26, 1961, ex. *Cinara* sp. on *Picea* sp., 3 ♀ (E. I. SCHLINGER).

Holotype ♀: England. B. M. Type Hym. 3 c. 89.

Hosts: (Unrevised literary data)

    Cinara laricis KOCH: FAHRINGER, 1937, North and Central Europe.

    Laricaria kochiana BOERNER: FAHRINGER, 1937, North and Central Europe.

Hosts: (Original and revised literary data).

    Buchneria pectinatae (NÖRDL.): STARÝ, 1962, on *Abies alba*, Czechoslovakia.

    *Cinara sp.: On *Picea* sp., Japan.

    Cupressobium juniperi (DEG.): STARÝ, 1960, on *Juniperus communis*, USSR – Eur. part.

Note: Mummified *Cinara* specimens are brown in color.

Pauesia pini (HALIDAY)

Aphidius (Aphidius) pini HALIDAY 1834, Ent. Mag. 2:96 (♀♂, Sweden, England). THOMSON, 1895. Opusc. Entomol. 20:2334 (♀♂, Sweden). MARSHALL, 1896, in ANDRÉ, Spec. Hym. Eur. et d'Alg. 5:566 (♀♂, England, host). MARSHALL, 1899, Trans. ent. Soc. London 1899: 41 (♀♂, England, host). SCHIMITSCHEK, 1935, Zbl. Forstwes. 61:215 (Czechoslovakia, host). SEITNER, 1936, Zbl. Forstwes. 62:46 (bion. Austria, host). SCHIMITSCHEK, 1936, Z. ang. Ent. 22:564 (Czechoslovakia, Austria, host). FAHRINGER, 1937, Festschr. 60. Geb. E. Strand Riga 3:243 (♀♂, Europe, host; *Coelonotus*). WATANABE, 1941, Ins. Mats. 15: 110-1 (Japan, host). PONTIN, 1960, Ent. mon. Mag. 85:154 (?, bion., England, host). STARÝ, 1960, Acta Faun. Ent. Mus. Nat. Pragae 6:23-6 (♀♂, Czechoslovakia, Germany, host; *Paraphidius*). MACKAUER, 1961, Beitr. Ent. 11:108 (notes on the type). STARÝ, 1962, Acta Soc. Ent. Cechosl. 59: in 42-58 (ecol., Czechoslovakia, host).

Aphidius lachnivorus ASHMEAD 1906, Proc. U. S. Nat. Mus. 30:189 (♂, Japan, host). WATANABE, 1957, Ins. Mats. 21: 2 (notes on the type specimen synonymy).

This species is easily distinguished from its close relatives by the sculpturing of tergite 1 and propodeum, and by the coloration.

Description – Female : Head transverse, shiny, smooth, sparsely haired, wider than thorax at tegulae, strongly and straightly narrowed behind eyes. Temple as wide as $1/3$ of transverse eye-diameter. Gena as wide as $1/4$ of longitudinal eye-diameter. Clypeus transverse, feebly shiny granulate, slightly arched and margined frontally, sparsely haired, separated by shallow groove from face. Tentorio-ocular line about $1/4$ shorter than intertentorial line. Eyes large, hemispherical, sparsely haired, strongly prominent laterally. Antennae 22-segmented, filiform, about as long as head, thorax and tergite 1 together. F_1 only a little longer than F_2.

Mesoscutum highly raised above prothorax and covering it when viewed laterally, slightly feebly granulate, sparsely haired. Notaulices distinct at the fore part, comparatively wide, somewhat rugose, feebly traced to nearly effaced on the disc. Propodeum (Fig. 54): Rami of the central carina strongly prominent both in transverse and longitudinal parts, completing large central pentagonal areola that differs by its declivity and concavity from its surrounding area; discs of areolae smooth, shiny, slightly granulate-rugose to granulate. Lateral areolae with sparse hairs. Wings (Fig. 164): Pterostigma triangular, strongly sclerotized, longer than metacarp. Radial abscissa 1 somewhat shorter or as long as width of pterostigma.

Abdomen lanceolate. Tergite 1 (Fig. 81) about 3 times as long as wide at spiracles, and twice as wide at apex as at spiracles, slender cephalad of spiracles, with shallow lateral impressions beyond spiracular tubercles and strongly dilating towards apex; this part being very flat, slightly impressed in the center; smooth, slightly rugose-granulate, with sparse long hairs. Spiracular tubercles somewhat prominent, situated at about the center of tergite. Following tergites smooth, shiny, sparsely haired. Genitalia as figured (Fig. 49).

Coloration variable. Head: frons, vertex, occiput more or less brownish-black. Temples, genae, face more or less yellowish rufous. Clypeus and mouthparts yellow. Antennae black; scape and pedicel especially in the lower part yellowish-rufous. Thorax: Prothorax yellowish-rufous to brown. Mesoscutum yellowish-rufous, more or less obscured on the lobes and in the neighborhood of praescutellar groove to entirely brownish-black. Scutellum yellowish-rufous to obscured. Remaining parts of thorax brownish-black. Wing venation brown. Pterostigma at least a little yellowed at base, cubital and anal veins colorless in the fore part. Fore legs yellowish-rufous, tarsi obscured. Middle and hind legs brown, coxae, femora partially,

base of tibiae and tarsi more or less yellowish-brown. Abdomen: Tergite
1 blackish-brown. Tergite 2 yellowish-rufous at base, the rest brown.
Following tergites with yellowish-rufous basal half and nearly brown
apex to entirely brown. Ovipositor sheaths brown.

Length of body about 3.5–4.2 mm.

Male: Antennae 23 to 24-segmented, reaching to about half of abdomen.
Tergite 1 more parallel-sided than in female, about twice as long as wide
at spiracles. Coloration: Head black, clypeus, mandibles (except apices) and
palpi yellowish-brown to brown. Thorax black. Wing-venation as in
female, only the white color of pterostigma is less distributed. Abdomen
black. Fore legs brown, tibiae and tarsi obscured. Middle and hind legs
blackish-brown; coxae, part of femora and base of tibiae lighter, otherwise
as described for female except for usual sexual differences.

General distribution: Europe and Japan.

Material examined: Japan: Nikko, ex. *Lachnus* sp., on *Larix* sp., 1♂, type
No. 7269, USNM Noboribetsu, May 19, 1961, ex. *Cinara laricis*, on *Larix* sp.,
139 ♂♀ (E. I. SCHLINGER).

Type specimens:

Aphidius pini HALIDAY: ♀, England. – B. M. Type Hym. 3. c. 90.

Aphidius lachnivorus ASHMEAD: ♂, Nikko, Japan, Cat. No. 7269, USNM.

Habitat: All specimens were collected in a young, dense larch forest.
Although a large series of specimens were collected, the parasite mummies
were not common (about one mummy per branch), but the host aphid was
extremely uncommon.

Hosts: (Unrevised literary data).

Cinara cembrae (CHOL.): SEITNER, 1936, on *Pinus cembra*, Austria.

Cinara laricicolus (MATS.): WATANABE, 1957, Japan.

Cinara laricis (WALK.): WATANABE, 1940, Japan.

Cinara nuda (MORDV.): FAHRINGER, 1937, Japan.

Cinara pini (L.): MARSHALL, 1896, 1899, England. SCHIMITSCHEK, 1936,
Austria. FAHRINGER, 1937, Europe. WATANABE, 1941, on *Pinus densiflora*, Japan.

Cinara sp.: MARSHALL, 1896, 1899, on *Pinus silvestris*, England.

Cinaropsis cistata (BCKT.): FAHRINGER, 1937, Europe.

Hosts: (Original and revised literary data).

Cinara laricis (WALK.): On *Larix* sp., Japan.

Cinara nuda (MORDV.): STARÝ, 1960, on *Pinus silvestris*, Germany.

Lachnidae sp.: STARÝ, 1960, on *Pinus silvestris*, Germany, Czechoslo-vakia.

Lachnidae sp. (probably *Cinara laricicolus* (MATS.)): see WATANABE, 1957, on *Larix* sp., Japan.

Note: Mummified aphids are black.

Pauesia tropicalis STARÝ and SCHLINGER, n. sp.

This species differs from other Far East Asian species of *Pauesia* in having the antennae more than 24-segmented in the female. It is related to the European *P. grossa* (FAHR.), but differs from the latter by the less prominent mesoscutum, the sculpturing of the propodeum, and by the shape of tergite 1.

Description – Female: Head transverse, strongly directly narrowed behind eyes, smooth, densely haired, as wide as thorax at tegulae. Occiput margin-ed. Temple $1/3$ shorter than transverse eye-diameter. Gena equal to half of longitudinal eye-diameter. Eyes large, nearly hemispherical, a little conver-gent towards clypeus, with sparse short hairs. Interocular line $1/6$ longer than transfacial line, equal to facial line. Clypeus with about 20 long hairs. Tentorio-ocular line $1/3$ shorter than intertentorial line. Antennae more than 24-segmented (broken), filiform, reaching nearly half of abdomen. F_1 equal to F_2, $1/3$ longer than wide. Socket-ocular line equal to socket-diameter.

Mesoscutum somewhat raised above prothorax, without covering it when viewed laterally; densely haired except for longitudinal places on lateral lobes. Notaulices coarsely rugose at the ascendent part and prolon-gately, more feebly rugose, along their length up to praescutellar groove. Propodeum (Fig. 52) with rather wide areola, with dense long hairs, coarsely rugose. Wing: Pterostigma triangular, little more than twice as long as wide. Metacarp little longer than pterostigma. Radial abscissa 1 longer than width of pterostigma (18:15), abscissa 2 shorter than 1 (14:18).

Abdomen lanceolate. Tergite 1 (Fig. 83) twice as long as wide at spiracles, gradually dilating towards apex, strongly prolongately rugose, with central prominent longitudinal carina. Spiracular tubercles hardly visible. Width at apex $1/2$ longer than at spiracles. Following tergites smooth, shiny, more densely haired towards apex. Ovipositor sheaths short, slightly arched upwards.

Coloration: Head dark brown, lower part of clypeus and mouthparts yellow. Antennae dark brown, scape on lower side yellow. Thorax blackish-brown, pronotum lighter. Wings hyaline, venation brown, cubital vein colorless at the fore part. Legs yellowish-brown, tarsi obscured. Tergite 1 blackish-brown, following tergites dark brown at the center and yellowish-brown at sides. Ovipositor sheaths brown.

Length about 5 mm.

Male: Unknown.

General distribution: Japan.

Holotype ♀: Japan, Kagoshima, April 11, 1961, ex. *Lachnus tropicalis* on *Ficus* sp., (E. I. SCHLINGER).

Habitat: What was apparently this species (or possibly *P. japonica* ASH.) was encountered several times near Kagoshima and also in Fukuoka, but except for one time, only emerged mummies were seen. The mummies, which are blackish brown in color, were seen in large numbers, often as many as 50 at a time. The habitat always seemed to be one of a forest condition.

Note: WATANABE (1939) bred the aphidiid species *P. japonica* (ASHM.) also from *Lachnus tropicalis* in Japan. The latter species is distinguishable (judging from the original description), by the 22-segmented antennae in the female, by the effaced notaulices on the disc of mesoscutum, and other features.

Pauesia unilachni (GAHAN)

Aphidius unilachni GAHAN, 1926, Proc. U. S. Nat. Mus. 70:1-2 (♀♂, Taiwan, host).

Aphidius praevisus GAUTIER and BONNAMOUR, 1936, Bull. Soc. Linn. Lyon N. S. 5:74-5 (♂♀, France, host). FAHRINGER, 1937, Festschr. 60. Geb. E. Strand, Riga 3:245 (notes). STARÝ, 1960, Acta Faun. Ent. Mus. Nat. Pragae 6:27, 29 (♀♂, Czechoslovakia, Germany, Sweden, host, *Paraphidius*). STARÝ, 1962, Acta Soc. Ent. Czechosl. 59: in 42-58 (ecol., Czechoslovakia, hosts).

This species is easily distinguished from its congeners by the number of antennal segments, sculpture of propodeum, shape of tergite 1 and the female genitalia.

Description – Female: Head transverse, strongly narrowed behind eyes, smooth, shiny, nearly $1/3$ wider than thorax at tegulae. Occiput margined.

Temple nearly as wide as transverse eye-diameter. Gena a little longer than $1/_3$ of longitudinal eye-diameter. Eyes large, widely oval, strongly prominent, sparsely shortly haired, slightly convergent towards clypeus. Interocular line $1/_4$ longer than transfacial line, a little longer than facial line. Clypeus slightly convex, separated by shallow groove from face, with about 8 to 12 long hairs. Tentorio-ocular line as long as $2/_3$ of intertentorial line. Antennae 16 to 17 (rarely 18)-segmented, filiform, reaching apex of tergite 1. F_1 2.5 times as long as wide at apex, a little longer than F_2. Socket-ocular line equal to socket-diameter.

Mesoscutum slightly elevated above prothorax, without covering it when viewed laterally. Notaulices deep, wide, crenulate at the ascendent part, effaced on the disc. Base of central lobe, lateral lobes along margins and effaced notaulices with sparse long hairs. Propodeum (Fig. 85) with very wide central areola, carinae strongly developed. Wing (Fig. 166): Pterostigma strongly sclerotized, 3 times as long as wide. Metacarp $1/_5$ shorter than pterostigma. Radial abscissa 1 slightly arched, less than one-half longer than abscissa 2.

Abdomen lanceolate. Tergite 1 (Fig. 82) about 3.5 times as long as wide at spiracles, about half wider at apex than at spiracles; rugose, with short central carina, sparsely haired. Spiracular tubercles strongly prominent, situated before center of tergite. Genitalia as figured (Fig. 45). Ovipositor sheaths very slender, curved upwards.

Coloration: Rather variable. Head brown to brownish-black; clypeus yellow to brown, mouthparts yellow except apices of mandibles. Antennae brown, often scape and pedicel yellowed at lower part. Thorax brown to brownish-black, pronotum, part of mesopleurae, metapleurae and propodeum sometimes with yellow or yellowish-brown spots. Tegulae and wing venation brownish-yellow to brown. Legs yellow (except apical half of tarsi), with more obscured tinge on femora and tibiae. Tergite 1, brownish-yellow to brown, following tergites brownish-yellow to brown, obscured towards apex. Ovipositor sheaths brown.

Length of body about 1.9 to 2.9 mm.

Male: Antennae 19 to 20-segmented. Head and thorax brown to brownish-black. Palpi and mandibles (except apices) yellow. Antennae brown to blackish-brown, lower part of scape and pedicel yellow. Wing venation and tegulae brown. Legs brownish-yellow, with obscured tinge on coxae, trochanters, femora and tibiae; tarsi except basal half obscured. Abdomen:

Tergite 1 brownish-yellow, remaining tergites brown, otherwise like female except for usual sexual differences.

General distribution: Europe (France, Germany, Czechoslovakia and Sweden), Hong Kong, Taiwan and South Korea.

Material examined (22 ♂♀ specimens). Hong Kong: Kowloon, Feb. 28, 1961, ex. *Cinara* sp., on *Pinus* sp., 9 ♂♀ (E. I. SCHLINGER). Same locality, Feb. 25, 1961, ex. *Cinara formosana* on *Pinus* sp., 1 ♀ (E. I. SCHLINGER). Taiwan: Taihoku, Sept. 3, 1926, ex. *Eulachnus*, 2 ♂♂ (R. TAKAHASHI, in USNM). South Korea: Seoul, May 8, 1961, ex. *Cinara orientalis*, on *Pinus* sp., 10 ♂♀ (E. I. SCHLINGER).

Type locality: Taiwan, Taihoku, ♀, ex. *Unilachnus* sp., Cat. No. 28983 USNM.

Habitat: This species was not uncommon in Far East Asia, but although it was located many times, usually the parasites had previously emerged. The species was always associated with *Cinara* sp. on *Pinus* sp. in definite forest habitats.

Hosts: (Unrevised literary data).

Aphidae sp.: GAUTIER & BONNAMOUR, 1936, on *Pinus* sp., France
Hosts: (Original and revised literary data).
  *Cinara orientalis TAK: on *Pinus* sp., S. Korea.
  *Cinara formosana TAK: on *Pinus* sp., Hong Kong.
  *Cinara sp.: on *Pinus* sp., Hong Kong.
  *Unilachnus sp.: GAHAN, 1926, Taiwan.
  Schizolachnus pineti (F.): STARÝ, 1960, *Pinus uliginosa*,
     Pinus silvestris, Czechoslovakia, *Pinus silvestris*, Germany.

Host specificity: Oligophagous, but primarily restricted to species of the genus *Schizolachnus* and related lachnids.

Species of *Pauesia* QUILIS from Japan and Taiwan unknown to the authors.

The following five species of *Pauesia* are at present unknown to the authors because of a paucity of specimens. However it seems advisable for completeness to quote the original description of these species.

Pauesia inouyei (WATANABE)

Aphidius inouyei WATANABE, 1941, Ins. Mats. 15:107-8 (♀♂, Japan, host).

"Female. – Head and thorax dark brown to black; face and mouth-parts fusco-testaceous, antennae dark brown, the two basal joints fusco-testaceous. Prothorax more or less fusco-testaceous. Wings slightly in-fumated, especially towards the apex, below the stigma and along the basal nervure; tegulae fusco-testaceous; stigma and veins brown, the intercubitus colorless. Legs fusco-testaceous, the middle pair more or less infuscate in the femora and tibiae; hind legs still darker, the femora at the apex, the tibiae except the base and the tarsi fuscous. Abdomen fusco-testaceous, the basal tergites fuscous, paler on each suture. Ovipositor-sheaths black.

Head transverse, smooth and shining, with white pubescence; antennae shorter than the body, 22- to 24-jointed. Thorax almost smooth and shining; mesonotum more or less rugulose, the parapsidal furrows obsolete. Pro-podeum almost smooth and shining, hollowed out at the base of 1st abdomi-nal segment, the excavation bordered in front and at the sides by sharp carinae, the median transverse carina forming the anterior border of the excavation, and the median longitudinal carina extending from the anterior margin to the transverse carina. Stigma 3 times as long as broad; metacarpus a little longer than the length of the stigma; 1st abscissa of the radius curved inwardly, as long as the stub of the 2nd, which is 3 times the length of the intercubitus; nervulus slightly postfurcal; nervus paralellus inter-stitial with the medial nervure. Abdomen lanceolate, longer than the head and thorax united; 1st tergite weakly rugulose, deeply excavated basally, linear to beyond the tubercles, which are situated at the middle, from the tubercles the lateral sides gradually widened towards the apex, 2.5 times as long as broad at the apex; 2nd and following tergites smooth and shining; ovipositor slightly exserted, the sheaths compressed, stout, somewhat curved upwards at the apex. Length, 3.5-4 mm.

Male. Closely resembles the female in general structure and color, but differs from the latter as follows: Antennae 25- or 26-jointed; 1st tergite stout, not so apparently widened towards the apex as in the female. Length, 3-3.5 mm.

Host: Cinara todocolus (INOUYE).

This species has been reared at Nopporo by Mr. M. INOUYE from *Cinara todocolus* infesting *Abies sachalinensis*.

Holotype (♀) and Allotype (♂): Nopporo, 26. VII. 1938, M. INOUYE leg., and deposited in the Entomological Institute of the Hokkaido Imperial University, Sapporo. Paratypes: 4 ♀♀, 1 ♂, 22. VI. 1937, 4 ♀♀, 4 ♂♂, IX. 1937, 9 ♀♀, 3 ♂♂, 26. VI. 1938, Nopporo, M. INOUYE leg.

Habitat: Hokkaido (Nopporo).

Remarks: This species comes near *Aphidius pictus* HALIDAY, from which it is easily distinguished by the structure of the 1st tergite, by the color of the thorax and by transparency of the wings." (original description).

Pauesia japonica (ASHMEAD)

Aphidius japonicus ASHMEAD, 1906, Proc. U. S. Nat. Mus. 30:189 (♀♂, Japan).

WATANABE, 1939, Ins. Mats. 13:83-4 (♀♂, Japan, host). WATANABE, 1957, Ins. Mats. 21:2 (notes on types).

"Female. – Length 4.8 mm. Head, sutures of scutellum, the metathorax and abdomen, except the petiole beneath and the apex of the second segment, black and shining, rest of thorax and the legs, except the two last joints of tarsi which are dusky, yellow. The antennae are long, filiform, 22-jointed, the first two joints more or less yellowish, the following joints black or brown black; joints 6 to 21 only about twice as long as thick, joints 3 to 5 a little longer. Wings hyaline, the stigma and veins light brownish, the basal nervure blackish.

Male. – Length 4 mm. Agrees well with the female, except that the mesothoracic lobes are sometimes dusky, the antennae longer, 24-jointed, while the abdomen beneath, the basal half of the third dorsal segment, and sometimes the sutures 4 and 5 are yellow.

Type. – Cat. No. 7268, USNM.

Locality. – Gigu (Y. Nawa). One female and 3 male specimens bred from an Aphis." (after ASHMEAD, 1906).

WATANABE (1957) notes the following on the type-material: "*Type:* 1 ♀ (lectotype) "34 4 27" (=27-IV-01), Gifu, Kunugi", 1 ♂ (paralectotype) "34 5 21" (=21-V-01), Gifu, Kunugi"; 2 ♂♂ (paralectotypes), "34 4 7 (=7-IV-01), Gifu, Yanagi.

Having examined the types, I have found that there are two different species among the paralectotypes: the one labeled "Kunugi", of which the host is probably *Pterochlorus tropicalis* V.D.G., is surely *Aphidius japonicus*, while the other two labeled "Yanagi", of which the host is probably *Tuberolachnus salignae* GMELIN, are identical with *Aphidius salignae* WATANABE (1957)."

Note: We have received through the kindness of C. F. W. MUESEBECK one male specimen from which we can support WATANABE's opinion in this matter.

Pauesia jezoensis (WATANABE)

Aphidius jezoensis WATANABE 1941, Ins. Mats. 15:108-9 (♀♂, Japan, host)

"*Female.* – Head and thorax dark brown; face, mouth-parts, two basal joints of the antennae and prothorax always fusco-testaceous; mesothorax sometimes fusco-testaceous beneath. Abdomen fusco-testaceous, somewhat infuscated dorsally. Wings hyaline; tegulae fusco-testaceous; stigma and veins fuscous. Legs fusco-testaceous; hind pair slightly infuscate. Ovipositor sheaths black.

Head transverse, smooth and shining, with white pubescence; antennae shorter than the body 18- or 19-jointed. Thorax smooth and shining; mesonotum more or less punctate, the parapsidal furrows slightly impressed posteriorly. Propodeum smooth and shining, somewhat rugose laterally, hollowed out at the base of the 1st abdominal segment, the excavation narrower than that of the preceding species, *Aphidius inouyei*, bordered in front and at the sides by sharp carinae; the median longitudinal carina extending from the anterior margin to the transverse carina which forms the anterior border of the excavation. Stigma 2.5 times as long as broad; metacarpus a little shorter than the length of the stigma; 1st abscissa of the radius a little longer than the stub of the 2nd, which is nearly equal to the intercubitus in length; nervulus interstitial or slightly postfurcal; nervus parallelus interstitial with the medial nervure. Abdomen lanceolate, longer than the head and thorax united; 1st tergite slightly convex, closely reticulate-rugulose, deeply excavated at the base, more or less widened towards the apex, 3 times as long as broad at the apex, the tubercles scarcely discernible; 2nd and following tergites smooth and shining; ovipositor subexserted, the sheaths compressed, slender, rather acute at the apex, and straight. Length 2.5-3.5 mm.

Male: Similar to the female, but differs from the latter in the following respects: Body and legs darker in color; wings whitish hyaline; abdomen fuscous above, with a luteous patch in the middle; antennae more slender than in the female, 20- to 22-jointed; 1st tergite fusco-testaceous, not at all widened towards the apex. Length, 2-3 mm.

Host: *Cinara noppororensis* INOUYE; *Lachniella costata* (ZETT.).

This species has been reared at Sapporo by Dr. H. KÔNO from *Cinara nopporensis* infesting *Picea jezoensis* and *Picea Glehni* and at Nopporo by Mr. M. INOUYE from the same aphid. It has been also reared at Sapporo by Dr. H. KÔNO from *Lachniella costata* infesting *Picea Glehni* and *Picea jezoensis*.

Holotype (♀): Sapporo, 6. VII. 1937, and Allotype (♂): Sapporo, 10. VII. 1937, reared from *Cinara nopporensis* by Dr. H. KÔNO, and deposited in the Entomological Institute of the Hokkaido Imperial University, Sapporo: Paratypes: 1♂, 25. VI. 1937, 1♂, 30. VI. 1937, 3 ♀♀, 2♂♂, 3. VII, 1937, Sapporo, reared from *Lachniella costata* by H. KÔNO; 1♂, 27. VI. 1937, 2 ♀♀, 6. VII. 1937, 1♂, 10. VII. 1937, 1♂, 10. VII. 1937, 1♂, 12. VII. 1937, Sapporo, reared from *Cinara nopporensis* by H. KÔNO; 2 ♀♀, 5 ♂♂, 27. VII. 1938, Nopporo, reared from *Cinara nopporensis* by M. INOUYE.

Habitat: Hokkaido (Sapporo and Nopporo).

Remarks: This species closely resembles the preceding species, *Aphidius inouyei*, but is easily distinguished from the latter by the number of antennal joints, by the venation and transparency of the wings, by the carination of the propodeum and by the structure of the 1st tergite and the ovipositor-sheath.''

Pauesia kônoi (WATANABE)

Aphidius kônoi WATANABE 1941, Ins. Mats. 15:106-7 (♂, Japan, host).

"*Male:* Black; mouth-parts and prothorax beneath fusco-testaceous; antennae dark brown. Wings hyaline, slightly infumated towards the apex; tegulae fusco-testaceous; stigma and vein brown. Legs fusco-testaceous; hind coxae fuscous basally. Abdomen black, with the 1st and 2nd sutures and the lateral margins of the 2nd tergite pale brown.

Head transverse, smooth and shining, with white pubescence; antennae longer than the body, 31-jointed. Thorax smooth and shining; parapsidal furrows strongly impressed, almost smooth. Propodeum reticulate – regulose, declivous at the strong transverse carina, the anterior face with median longitudinal carina which extends from the anterior margin to the transverse carina, and the posterior face broadly excavated medially, with two widely separated longitudinal carinae extending from the transverse carinae to the posterior margin and bordering the excavation. The outlines of the stigma and nervures in the apical half of the fore wing weakly defined, not so sharp as those in the basal half. Stigma 3 times as long as broad; 1st abscissa of the radius as long as the stub of the 2nd; intercubitus two-thirds the length of the 2nd abscissa of the radius; nervulus postfurcal

by its own length. Abdomen clavate, a little longer than the head and thorax united; 1st tergite convex, finely rugulose, linear to beyond the tubercles which are situated anterior to the middle, from the tubercles the lateral
sides gradually widened towards the apex, 2 times as long as broad at the
apex; 2nd and following tergites smooth and shining. Length, 4.5-5 mm.

Female: Unknown

Host: *Cinara longipennis* (MATSUMURA).

This species has been reared at Sapporo by Dr. H. KÔNO from *Cinara
longipennis* infesting *Abies sachalinensis*.

Holotype (♂): Sapporo, 2. VII. 1937, H. KÔNO leg., deposited in the
Entomological Institute of the Hokkaido University, Sapporo. Paratypes:
1 ♂, Sapporo, 30. VI. 1937 and 1 ♂, Sapporo, 2. VII. 1937, H. KÔNO leg.

Habitat: Hokkaido (Sapporo).

Remarks: This species resembles *Aphidius grossus* (FAHRINGER), a parasite
of *Cinara cecconii* DEL GUERCIO in Europe, but is easily distinguishable from
the latter by the colour and by the structure of the propodeum." (original
description).

There is a recent note on this species by WATANABE & TAKADA (1964 b).

Pauesia laticeps (GAHAN)

Aphidius laticeps GAHAN 1926, Proc. U.S. Nat. Mus. 70 Art. 8, 2-3 (♀♂,
Taiwan, host)

"Resembles *A. pinaphidis* ASHMEAD, but the mesoscutum is less strongly
sculptured, the wing stigma is more triangular, and the ovipositor sheaths
are broader and shorter.

Female. – Length 3.8 mm. Head smooth, viewed from above broader than
the thorax, fully twice as broad as long; viewed from in front the head is obviously broader than high; eyes large and prominent, nearly circular, sparsely hairy; face twice as broad as high; palpi short; antennae broken, the
first flagellar joint about two and one-half times as long as thick, following
joints shorter; mesoscutum faintly alutaceous and subopaque, with a few
obscure wrinkles following the subobsolete notauli, the anterior one-third
with some distinct rugosities or sub-obsolete punctures; propodeum nearly
smooth above but with its posterior face and lateral margins distinctly rugulose, the petiolar area concave, much broader than long and more or less
transversely wrinkled within; the lateral areas of posterior face of propodeum small and mostly restricted to the lateral angles; pleura smooth; legs

normal; stigma of forewing short and broad, emitting the radius at the middle; metacarpus distinctly longer than the stigma; radius short, somewhat thickened basally, its first abscissa less than twice as long as the stub of second, which is a little longer than the intercubitus; brachial cell closed; abdomen one and one-half times as long as the thorax, the first tergite rugose, twice as broad at apex as at base; ovipositor sheaths rather broad and short. General color reddish testaceous; metanotum, propodeum, apical half or more of third, fourth, and fifth and all the following abdominal segments dark brownish to blackish; scape and pedicel testaceous, flagellum blackish; legs concolorous with thorax, the posterior femora and tibiae suffused with brownish; wings hyaline; venation brownish, the costal and basal veins darker than the others, and the median and submedian veins mostly pale; stigma at base and narrowly along the anterior margin pale, otherwise brownish.

Male. Unknown

Type locality. – Taihoku, Formosa.

Type. Cat. No. 28984, USNM.

Host. Dilachnus, species.

One female received from T. SHIRAKI and said to have been parasitic upon an unidentified species of *Dilachnus*, collected by R. TAKAHASHI. The type host has lost one pair of wings and both antennae are broken." (original description).

GENUS PRAON HALIDAY

Aphidius NEES subg. *Praon* HALIDAY 1833, Ent. Mag. 1:483

Type species: Bracon exoletus NEES.

Aphidaria PROVANCHER 1886, Add. Faun. Canad. Hym., p. 151.

Type species: Aphidaria simulans PROVANCHER.

Literary data: MARSHALL, 1896, in ANDRÉ, Spec. Hym. Eur. et d' Alg. 5: 532-3, etc. (European spp.). MARSHALL, 1899, Trans. ent. Soc. London 1899: 14, etc. (European spp.). SMITH, C. F., 1944, Ohio State Univ. Contr. Zoo. Ent. 6:22, etc. (Revision of Nearctic spp.). MACKAUER, 1959, Beitr. Ent. 9:810-65, etc. (Revision of Palearctic spp.). WATANABE & TAKADA, 1964a, Ins. Mats. 27(1): 8-11 (note on two Japanese *Praon*).

Description: Head subcubical, as wide as or wider than thorax at tegulae. Occiput margined. Antennae filiform, with variable number of segments (13 to 23). Mesoscutum strongly prominent, vertically falling to prothorax.

Notaulices distinct throughout. Propodeum smooth. Fore wing: Pterostigma triangular, slender at the fore part. Metacarp distinct. Radial vein developed, never reaching wing-apex so that pterostigmal cell is incomplete. Radial cell 1 and median cell 1 distinctly separated by median abscissa 1. Interradial veins effaced. Intermedian vein more or less distinct. Hind wing with complete basal cell. Tergite 1 quadrate or a little longer than wide. Abdomen of female lanceolate, rounded in male. Ovipositor sheaths nearly straight to slightly curved upwards, slenderly triangular, pointed at apex, with sparse hairs.

General distribution: Holarctic and Oriental regions.

Bionomics: Parasite of aphids with pupation occurring inside separate cocoon underneath the parasitized aphid.

Key to the species of *Praon* (♀♀)

1 Tergite 1 distinctly longer than wide at spiracles 2
  Tergite 1 quadrate   *Praon quadratum* STARÝ and SCHLINGER, n. sp.
2 Mesoscutum densely pubescent, lateral lobes sometimes with small oval hairless spots   *Praon orientale* STARÝ and SCHLINGER, n. sp.
  Mesoscutum sparsely pubescent, lateral lobes almost entirely hairless
                Praon glabrum STARÝ and SCHLINGER, n. sp.

Praon glabrum STARÝ and SCHLINGER, new species

This species is related to *P. orientale* n. sp. but is easily distinguished by the pubescence on the mesoscutum and head.

Description – Female: Head subcubical, smooth, shiny, wider than thorax at tegulae, sparsely haired (also on face). Temple equal to transverse eye-diameter. Gena equal to longitudinal eye-diameter. Eyes of medium size, oval, sparsely haired. Interocular line less than $1/2$ longer than transfacial line, somewhat shorter than facial line. Clypeus with some long hairs. Tentorio-ocular line equal to $1/5$ of intertentorial line. Antennae 17-segmented, filiform. F_1 5 times as long as wide, F_2 $1/4$ shorter than F_1. Socket-ocular line shorter than socket-diameter.

Mesoscutum falling vertically to prothorax. Central lobe densely pubescent, lateral lobes nearly hairless, only with sparse hairs along margins. Propodeum smooth, pubescent. Wing: (Fig. 170) Pterostigma 3 times as long as wide, metacarp $1/4$ shorter than pterostigma. Radial vein equal to pterostigma.

Abdomen lanceolate. Tergite 1 (Fig. 86) somewhat longer than wide at spiracles, slightly rugose to nearly smooth, slightly convex, with distinct lateral carinae. Distance between spiracular tubercles and apex equal to width at spiracles. Spiracular tubercles hardly prominent. Following tergites smooth, shiny, sparsely haired. Genitalia as figured (Fig. 47).

Coloration: Head dark brown; face, lower part of genae, clypeus and mouthparts yellowish-brown. Scape, pedicel and F_1 yellowish-brown, rest brown. Thorax yellowish-brown, apical part of mesoscutum and metanotum brown. Wings hyaline, venation brown. Legs yellow, apices of tarsi darkened. Tergite 1, 2 and base of 3 yellow, remaining tergites brown, last sternite yellow and ovipositor sheaths dark brown.

Length of body about 2.1 mm.

Male: Antennae 20-segmented. Scape and pedicel brown. Thorax with more or less brown coloration. Apex of abdomen brown. Otherwise as described for the female except for sexual differences.

General distribution: Japan.

Holotype ♀: Sapporo, Japan, May 24, 1961, ex. *Euceraphis betulae* on *Betula* sp. (E. I. SCHLINGER). Allotype and 1 ♂ paratype are topotypical.

Habitat: The type specimens were collected on an ornamental planting of *Betula* in the Botanical Garden in Sapporo.

Note: Parasite cocoon is white and mummified aphid is light brown.

Praon orientale STARÝ and SCHLINGER, new species

This species differs from other Far East Asian species of *Praon* by the shape of tergite 1, by the pubescence of the mesoscutum and by the coloration of the abdomen. It seems to be related to the European *P. volucre* (HAL.), but differs from it easily by coloration alone.

Description – Female: Head subcubical, rounded, smooth, shiny, wider than thorax. Temple about equal to transverse eye-diameter. Gena equal to $^1/_5$ or $^1/_6$ of longitudinal eye-diameter. Eyes of medium size, oval, sparsely and shortly haired, convergent towards clypeus. Face densely haired, the rest of head with sparse hairs. Interocular line a little more than $^1/_2$ longer than transfacial line, shorter than facial line. Clypeus densely haired. Tentorio-ocular line about $^1/_5$ of intertentorial line. Antennae 17 to 20-segmented, filiform, reaching to half of abdomen. F_1 5 times as long as wide, F_2 about $^1/_5$ shorter than F_1. Socket-ocular line shorter than socket-diameter.

Mesoscutum falling vertically to prothorax. Central lobe densely pubescent, lateral lobes sometimes with small hairless oval spots. Notaulices distinct throughout. Propodeum smooth, pubescent. Wing (Fig. 169): Ptero-stigma about 4 times as long as wide, about $^1/_2$ longer than metacarp. Radial vein a little longer than metacarp. Median abscissa 1 partly colorless, but intermedian vein colored.

Abdomen lanceolate. Tergite 1 (Fig. 87) $^1/_3$ longer than wide at spiracles, more or less prolongately rugose (sculpturing variable), sparsely haired, more or less concave, with more or less distinct central carina near the base and with lateral longitudinal carinae. Distance between spiracular tubercles and apex equal to width at spiracles, or somewhat longer. Spiracular tubercles strongly prominent. Following tergites smooth, shiny and sparsely haired, genitalia as figured (Fig. 44).

Coloration variable: Head dark brown; clypeus, genae, sometimes also face yellow to yellowish-brown, mouthparts yellow. Scape, pedicel, F_1 and sometimes base of F_2 yellow, the rest brown. Thorax with variable colo-ration. Usually mesoscutum and metanotum brown, the rest yellowish-brown or thorax mostly dark brown to brown with lighter prothorax. Wings hyaline, venation brown. Legs yellow, apices of tarsi darkened. Ter-gite 1 yellowish-brown, tergite 2 yellow, remaining tergites dark brown except apical ones which are lighter colored. Apical sternites yellow. Ovi-positor sheaths dark brown.

Length of body about 2.3 to 2.5 mm.

Male: Antennae 21 to 23-segmented. Scape and pedicel yellow, flagellar segments brown. Abdomen entirely brown at apex. Otherwise like the female except for sexual difference.

General distribution: Hong Kong, South Korea, Japan and Taiwan.

Holotype ♀: Hong Kong, Sha Tin, February 28, 1961, ex. *Macrosiphum rosaeibarae* on *Rosa* sp. (E. I. SCHLINGER). Allotype ♂ and 14 ♂♀ paratypes are topotypical.

Other specimens examined were 93 paratypes as follows: Hong Kong: Kowloon, February, 1961, ex. *Myzus* sp. on *Citrus* sp. 1 ♀ (E. I. SCHLINGER). South Korea: Cheju-do Island, May 6, 1961, ex. *Macrosiphum rosaeibarae* on *Rosa* sp., 6 ♂♀ (E. I. SCHLINGER); Cheju-do Island, May 6, 1961, ex. *Cavariella* sp. on *Rosa* sp. 1 ♂ (E. I. SCHLINGER); Cheju-do Island, May 6, 1961, ex. *Myzus persicae* on unknown plant, 1 ♂, 1 ♀ (E. I. SCHLINGER); Seoul, May 9, 1961, ex. *Aphis spiraecola* on *Crataegus pinnatigida*, 1 ♂, 1 ♀ (E. I. SCHLINGER). Japan: Fukuoka, April 19, 1961, ex. *Amphorophora magnoliae* on *Sambucus* p.,7 ♂♀ (E. I. SCHLINGER); Fukuoka, April 18, 1961, ex. *Acyrthosiphon* sp. on

Convolvulus sp., 1 ♀ (E. I. SCHLINGER); Fukuoka, April 13, 1961, ex. *Aphis spiraecola* on *Spiraea thunbergi*, 4 ♂♀ (E. I. SCHLINGER); Sapporo, May 27, 1961, ex. *Acyrthosiphon* sp. on *Syringa* sp. 15 ♂♀ (E. I. SCHLINGER); Sapporo, May 24, 1961, ex. *Acyrthosiphon* sp. on *Symphoricarpus mollis*, 5 ♂♀ (E. I. SCHLINGER); Sapporo, May 24, 1961, ex. *Acyrthosiphon* sp. on *Quercus* sp. 1 ♂, 1 ♀ (E. I. SCHLINGER); Sapporo, May 27, 1961, ex. *Acyrthosiphon* sp. on *Mentha sacha-liensis*, 1 ♂ (E. I. SCHLINGER); Sapporo, May 24, 1961, ex. *Amphorophora magnoliae*, on *Sambucus sieboldiana*, 23 ♂♀ (E. I. SCHLINGER); Kagoshima, April 12, 1961, ex. *Acyrthosiphon* sp. on *Aralia* sp., 1 ♂, 1 ♀ (E. I. SCHLINGER); Kagoshima, April 12, 1961, ex. *Amphorophora magnoliae* on *Sambucus* sp., 4 ♂♀ (E. I. SCHLINGER); Asakawa, April 27, 1961, ex. *Parachaitophorus spi-raeae* on *Spiraea nervosa*, 1 ♀ (E. I. SCHLINGER); Asakawa, May 12, 1961, ex *Rhopalosiphoninus tiliae* on *Tilia japonica*, 4 ♂♀ (E. I. SCHLINGER); Asakawa, April 27, 1961, ex. *Rhopalosiphoninus deutzifoliae* on *Deutzia crenata*, 5 ♂♀ (E. I. SCHLINGER); Asakawa, May 12, 1961, ex. *Acyrthosiphon* sp. on *Paraben-zon praecox*, 1 ♀ (E. I. SCHLINGER); Asakawa, May 12, 1961, ex. *Amphorophora* sp. on *Philadelphus satsumanus*, 1 ♀ (E. I. SCHLINGER); Tokyo, May 13, 1961, ex. *Macrosiphum rosaeibarae* on *Rosa multiflora*, 4 ♂♀ (E. I. SCHLINGER). Taiwan: Wushe (=Musha), March 12, 1961, ex. *Hyperomyzus lactucae* on *Lactuca sativa*, 1 ♂ (E. I. SCHLINGER).

Habitat: This parasite does not seem to occupy any particular habitat, but it was most often encountered in arboreal situations in forest-like areas.

Hosts:

*Acyrthosiphon spp.: On *Parabenzon praecox*, Japan.

On *Quercus* sp., Japan. On *Convolvulus* sp., Japan. On *Syringa* sp., Japan. On *Symphoricarpus mollis*, Japan. On *Aralia* sp., Japan. On *Men-tha sachaliensis*, Japan.

*Amphorophora magnoliae (ESSIG and KUWANA): On *Sambucus* sp., Japan. On *Sambucus sieboldiana*, Japan.

*Amphorophora sp.: On *Philadelphus satsumanus*, Japan.

*Aphis spiraecola PATCH: On *Crataegus pinnatigida*, South Korea. On *Spi-raea thunbergi*, Japan.

*Cavariella sp.: On *Rosa* sp., South Korea.

*Hyperomyzus lactucae (L.): On *Lactuca sativa*, Taiwan.

*Macrosiphum rosaeibarae MATS.: On *Rosa* sp., Hong Kong and South Korea.

On *Rosa multiflora*, Japan.

*Myzus persicae SULZ.: South Korea.

Myzus sp.: On *Citrus* sp., Hong Kong.

Parachaitophorus spiraeae TAKAH: On *Spiraea nervosa*, Japan.

Rhopalosiphoninus tiliae MATS.: On *Tilia japonica*, Japan.

Rhopalosiphoninus deutzifoliae SHINJI: On *Deutzia crenata*, Japan.

Host-specificity: Although this species appears to be widely polyphagous on aphids occurring in forest-type habitats, it seems to prefer aphids in the *Macrosiphina* group being particularly common on species of *Acyrthosiphon*.

Note: The cocoons of *P. orientale* are white and the mummified aphids are light brown.

Praon quadratum STARÝ and SCHLINGER, new species

This species is easily distinguishable from its congeners by the quadrate shape of tergite 1.

Description – Female. Head subcubical, rounded, somewhat wider than thorax at tegulae. Temple equal to transverse eye-diameter. Gena equal to $1/6$ of longitudinal eye-diameter. Eyes of medium size, oval, sparsely shortly haired. Interocular line equal to nearly twice of transfacial line, somewhat shorter than facial line. Face pubescent, otherwise head sparsely haired. Clypeus with long dense hairs. Tentorio-ocular line equal to $1/5$ of intertentorial line. Antennae (partially broken); F_1 5 times as long as wide, $1/4$ longer than F_2. Socket-ocular line shorter than socket-diameter.

Mesoscutum falling vertically to prothorax, sparsely and long haired, lateral lobes with small hairless spots. Notaulices distinct throughout. Propodeum smooth, pubescent. Wing: Pterostigma 4.5 times as long as wide, somewhat longer than metacarpus. Radial vein equal to metacarp.

Abdomen lanceolate. Tergite 1 (Fig. 88) square, almost smooth, slightly convex. Distance between spiracular tubercles and apex shorter than width at spiracles. Spiracular tubercles poorly prominent. Remaining tergites smooth, shiny, sparsely haired.

Coloration: Head dark brown, face lighter colored. Lower part of genae and clypeus yellowish-brown. Mouthparts yellowish-brown to yellow. Scape, pedicel and the greatest part of F_1 yellow, the rest of antennae brown. Thorax dark brown, prothorax yellowish-brown. Wings hyaline, venation brown. Legs yellow. Tergite 1, 2 mostly, and base of 3 yellow, rest of abdomen brown. Last sternite light brown. Ovipositor sheaths dark brown.

Length of body about 1.5 mm.

Male: Unknown.

Holotype ♀: Japan, Osaka, April 21, 1961, ex. *Coloradoa rufomaculata* on *Chrysanthemum* sp. (E. I. SCHLINGER).

Habitat: The holotype was collected in a *Chrysanthemum* garden.

Note: The cocoon is white and the mummified aphid is very light brown.

Undetermined species of *Praon* HALIDAY

We have been able to differentiate at least two more Far East Asian *Praon* species which perhaps represent new species. However, lack of specimens prevents us from forming any conclusions at this time, except to give certain data which may be helpful to the next reviser.

Praon species – No. 1

Female: Antennae 16-segmented. Mesoscutum with dense long hairs. Tergite 1 slender, longer than width at spiracles, nearly smooth and convex.

Length of body about 1.4 mm.

Coloration: Head brown, clypeus and mouthparts yellow, scape yellow, pedicel obscured, F_1 yellow, the rest of antennae brown. Thorax brownish-yellow. Legs yellow, apices of tarsi darkened. Tergite 1, base of 2 and little of last sternite yellowish–brown, rest of abdomen brown. Ovipositor sheaths dark brown.

Material examined: Japan: Kagoshima, April 11, 1961, ex. aphid on *Woodwardia* sp., 1 ♀, (E. I. SCHLINGER).

Habitat: The unique specimen of this species was collected on a humid, shrub–covered hillside close to the ocean. No live aphids were available and several other cocoons found were previously emerged.

Note: The cocoon is white and the mummified aphid is light brown.

Praon species – No. 2

Male: Antennae 21-segmented. Mesoscutum with long dense hairs. Propodeum with comparative sparse long hairs. Tergite 1 nearly smooth, longer than width at spiracles, slightly convex, width at spiracles equal to their distance from apex.

Coloration: Head brown, mouthparts, lower part of genae and clypeus yellowish-brown. Scape and pedicel yellow, the rest brown. Thorax brown. Legs yellowish-brown, tarsi obscured. Abdomen yellow, tergites with lightly obscured margins (immature specimen?).

Length of body about 1.4 mm.

Material examined: South Korea: Cheju-do Island, May 6, 1961, ex. *Aphis gossypii* on unknown plant, 1 ♂ (E. I. SCHLINGER).

Praon species

We are also including some records of various swept specimens of *Praon* recently collected from the Far Eastern part of the USSR.

Material examined: USSR Primorye: Buchta Shamora, env. of Vladivostok, July 28, 1961. (TRIAPICYN). Suputinsky zapovednik, southern slope of mixed wood, August 8, 1961, (SCHUVAKHINA). Gorno-Taiozhnaja, edge of wood, July 10, 1961 (TRIAPICYN).

GENUS PROTAPHIDIUS ASHMEAD

Coelonotus FÖRSTER 1862, Verh. Nat. Ver. Preuss. Rheinl. 19: 248 (Preoccupied)

   Type species: Coelonotus rufus FÖRSTER

Protaphidius ASHMEAD 1900, Canad. Ent. 32:368.

   Type species: Coelonotus rufus FÖRSTER

Menozzia GOIDANICH 1934, Boll. Lab. ent. Bologna 6:217-27.

   Type species: Menozzia formicaria GOIDANICH

   Literary data: STARÝ, 1958, Acta Faun. Ent. Mus. Nat. Pragae 3:89-93 (revision).

This genus is quite similar to *Pauesia* QUILIS. It differs from the latter by the shape of female abdomen, the female genitalia.

Description: Head transverse to slightly square, wider than thorax. Eyes hemispherical, strongly prominent laterally. Antennae 24 to 28 segmented, moniliform. Mesoscutum highly elevated above pronotum when viewed laterally. Notaulices distinct at the fore part or throughout all their length. Venation of wings partly reduced: Pterostigma large, triangular, strongly sclerotized, pterostigmal cell open, radial and median cells confluent, complete. Hind wing with complete basal cell. Propodeum strongly declivous backwards, divided with carinae on several smaller areolae. Abdomen of female behind third abdominal segment narrow, tubiform and telescopic. Ovipositor small, slightly curved downwards.

General distribution: Palearctic region (Europe and Japan).

Bionomics: Parasites of aphids. Pupation inside parasitized aphid.

Protaphidius nawaii (ASHMEAD)

Aclitus nawaii ASHMEAD 1906, Proc. U.S. Nat. Mus. 30: 188 (♀♂, Japan; host)

Differs from the European *P. wissmannii* (RATZEBURG) by the entirely punctate notaulices.

Description – Female : Head wider than thorax at tegulae, transverse to slightly square, strongly narrowed behind eyes. Temple about half as long as transverse eye–diameter. Gena a little shorter than longitudinal eye–diameter. Clypeus with deep tentorial pit on either side, not distinctly separated from face by two longitudinal impressions, distinctly margined frontally, slightly depressed in the middle part, with sparse long hairs. Eyes large, hemispherical, strongly prominent laterally. Ocelli large, hemispherical, forming a triangle slightly elevated over surface of vertex, a shallow groove directed to antennal bases runs from the fore ocellus. Antennae 24-segmented, moniliform, thick, pubescent, about as long as head, thorax and tergite 1 together; each segment with dark ring at apex.

Pronotum narrow in median region, slightly prominent before tegulae and there definitely longitudinally sculptured. Mesoscutum highly elevated above pronotum, with surface densely haired. Notaulices very distinct, deeply punctate and their environment strongly rugose; therefore mesoscutum in fore part is nearly of edged appearance as seen from above. Praescutellar groove deep and smooth. Mesopleurae smooth, with scattered hairs, foveately margined. Metanotum projecting in obtuse spine in the middle and deeply impressed laterally. Metapleurae strongly irregularly rugose, especially on margins, sparsely haired. Wing: Pterostigma large, triangular, strongly sclerotized, otherwise with characteristics of genus. Legs slender, relatively long, sparsely haired. Propodeum strongly declivous backwards, with two prominent protuberances in hind part; with central longitudinal carina archingly bifurcating in about the middle, both rami nearly reaching hind margin of propodeum; surface between rami transversely striated, strongly declivous backwards; with further longitudinal carinae running on either side, joining rami of central carina near spiracular tubercles, about half way between central carina and lateral margin of propodeum. Area between central carina and both lateral carinae regularly transversely striated similar to strongly declivous area between both rami of central carina. Area between lateral carinae and lateral margins of propodeum irregularly rugose; with few scattered hairs in rugose portion and around spiracles. Spiracular tubercles small, portion around spiracular tubercles slightly elevated in lateral view; there lateral carinae join rami of central carina.

Abdomen: Tergite 1 narrowly connected with propodeum, immediately dilating behind as far as spiracular tubercles which are situated on the first third of segment; from spiracular tubercles, tergite 1 progressively and slowly dilates to apex where it is widest; its surface irregularly longitudinally striated and sparsely haired. Spiracular tubercles distinctly prominent laterally. Segments 2 and 3 smooth and shining, fused, the latter relatively densely haired apically and strongly narrowed behind; following abdominal segments tubiform, telescopic, presenting a distinct, narrowed part of abdomen (Fig. 28). This tubiform part of abdomen about twice longer than preceding abdominal segments combined (in dried specimens a little shorter). Ovipositor very little visible, slightly curved downwards. Genitalia as in Fig. 28.

Coloration: Brown fuscous except for apex of antennae, mesopleurae, propodeum, coxae, and tergite 1 which are darkened. Wings transparent with dark spots.

Length of body about 8 mm.

Male: Antennae 28-segmented. Abdomen oval, normal, otherwise as described for the female except for sexual differences.

General distribution: Europe and Japan.

Material examined: Japan: Gifu, Oct. 1902, 1 ♀ (Y. NAWA). Type No. 9266 in USNM.

Habitat: Unknown.

Host: Stomaphis yanonis Tak: (WATANABE, 1957, Japan).

Note: Owing to lack of material all the figures are drawn from *P. wissmannii* (RATZ.) (Figs. 28, 53).

GENUS TRIOXYS HALIDAY

Aphidius NEES subg. *Trioxys* HALIDAY 1833, Ent. Mag. 1:261, 488

Type species: Aphidius cirsii CURTIS

Neuropenes PROVANCHER 1866, Addit. Corr. Faune Ent. Canada Hym., pp. 151, 153.

Type species: Neuropenes ovalis PROVANCHER

Subgenera

Betuloxys MACKAUER

 Trioxys HALIDAY subg. *Betuloxys* MACKAUER 1960, Beitr. Ent. 10:139

  Type: Trioxys compressicornis RUTHE

Binodoxys MACKAUER

 Trioxys HALIDAY subg. *Binodoxys* MACKAUER 1960, Beitr. Ent. 10:141

  Type species: Aphidius (Trioxys) angelicae HALIDAY

Fissicaudus STARÝ and SCHLINGER, new subgenus

 Trioxys HALIDAY subg. *Fissicaudus* n. subg.

  Type species: Trioxys (Binodoxys) confucius MACKAUER

Pectoxys MACKAUER

 Trioxys HALIDAY subg. *Pectoxys* MACKAUER 1960, Beitr. Ent. 10:154-5

  Type species: Trioxys (Trioxys) macroceratus MACKAUER

Trioxys s. str.

 Trioxys HALIDAY subg. *Trioxys* s. str., MACKAUER 1960, Beitr. Ent. 10:
 149

  Type species: Aphidius cirsii CURTIS

Literary data: MARSHALL, 1896, in ANDRÉ, Spec. Hym. Eur. et d'Alg. 5: 550-1 (Eur. spp.). MARSHALL, 1899, Trans. ent. Soc. London 1899: 26-7 (Eur. spp.). SMITH, 1944, Ohio State Univ. Contr. Zoo. Ent. 6:84-5 (revision of Nearctic spp.). MACKAUER, 1959, Beitr. Ent. 9:144-79 (rev. of Eur. spp.). MACKAUER, 1960, Beitr. Ent. 10:137-60 (rev. of Eur. spp.).

This genus is easily distinguishable from all the genera of Aphidiidae by the wing-venation and by having two prongs on the last abdominal sternite in the females.

Description: Head transverse. Antennae filiform, ($\female$) 10 to 12, ($\male$) 12 to 14-segmented. Notaulices distinct at the ascendent part of mesoscutum. Propodeum smooth or areolated. Fore wing: Pterostigma triangular. Radial vein distinctly developed. Pterostigmal cell open. Cubital cell 2 more or less distinct, otherwise the venation effaced behind the basal vein towards wing-apex. Hind wing with more or less distinct basal cell. Tergite 1 with primary (=spiracular) and secondary or only with primary tubercles, with variable sculpturing. Ovipositor sheaths curved downwards. Last sternite of females with a pair of simple prongs.

General distribution: Nearly cosmopolitan.

Bionomics: Parasite of aphids with pupation occurring in the host skin of the parasitized aphid.

Key to the species and subgenera of *Trioxys* (♀♀)

1   Tergite 1 with primary (=spiracular) and secondary tubercles (Fig. 93), the latter being sometimes hardly visible because of being nearly fused with primary tubercles (Fig. 97)................. 2

     Tergite 1 with primary tubercles only (Fig. 90) (*Trioxys* s. str.) ... 10

2(1) Prongs of last sternite beginning at apex of sternite (Fig. 13) – (*Binodoxys* MACKAUER) 3

     Prongs of last sternite beginning near base of sternite (Fig. 15) (*Fissicaudus* STARÝ and SCHLINGER n. subg.)

                             T. (F.) confucius MACKAUER

3(2) Distance between primary and secondary tubercles longer or equal to width at spiracles (Figs. 91) 4

     Distance between primary and secondary tubercles shorter than width at spiracles (Figs. 93)............................. 9

4(3) Interocular line twice (or nearly) as long as transfacial line. Prongs of last sternite with 6 to 8 long hairs on dorsal margin 5

     Interocular line $^1/_2$ longer than transfacial line. Prongs of last sternite with 4 to 5 long hairs.......................... 7

5(4) Head and thorax entirely brown, only neighborhood of mouth-parts and mouthparts yellow or yellowish-brown. Prongs with 6 long hairs on dorsal margin *T. (B.) brunnescens* STARÝ & SCHLINGER n. sp.

     Head and thorax with variously distributed orangish-yellow coloration. Prongs of last sternite with 6 to 8 long hairs........ 6

6(5) Tergite 1 with long and prominent central carina (Fig. 93). Eyes with dense long hairs. Clypeus with 7 to 8 long hairs

                T. (B.) carinatus STARÝ and SCHLINGER n. sp.

     Tergite 1 with feeble central carina (Fig. 94). Eyes with sparse short hairs. Clypeus with 4 to 5 long hairs

             T. (B.) orientalis STARÝ and SCHLINGER n. sp.

7(4) Head and thorax nearly entirely brown, only neighborhood of mouthparts and mouthparts and sometimes prothorax and propodeum yellowish brown. Prongs with 4 long hairs

             T. (B.) indicus SUBBA RAO and SHARMA

     Head and thorax mostly yellowish-orange. Prongs with 5 long hairs 8

8(7) Apical hairs on prongs unusually long, nearly as long as those on dorsal margin (Fig. 50) .... *T. (B.) struma* GAHAN

Apical hairs on prongs short (Fig. 19) *T. (B.) sinensis* MACKAUER

9(3) Primary and secondary tubercles distinct, the part between them distinctly distinguishable from basal and apical parts of tergite (Fig. 98) *T. (B.) "rietscheli"* MACKAUER group

Primary and secondary tubercles hardly distinguishable, secondary tubercles indicated by hairs, part between them hardly distinguishable from basal and apical parts (Fig. 47)

T. (B.) communis GAHAN

10(1) Metacarp very short, nearly point-like. Radial vein somewhat longer than $1/2$ of pterostigmal length (Fig. 187). Antennae 12-segmented. Propodeum smooth, with 2 divergent more or less visible carinae at lower part *T. (T.) asiaticus* TELENGA

Metacarp as long as $1/2$ of pterostigma. Radial vein about as long as pterostigma (Fig. 186). Antennae 11-segmented. Propodeum areolated *T. (T.) luteolus* STARÝ and SCHLINGER n. sp.

SUBGENUS *Binodoxys* MACKAUER

Description: Head transverse. Antennae filiform, (♀) 10 to 11, (♂), 12 to 13-segmented. Propodeum more or less areolated. Tergite 1 with primary and secondary tubercles, the latter sometimes hardly distinguishable because of their approximation to primary tubercles.

General distribution: Holarctic and Oriental regions.

Trioxys (Binodoxys) brunnescens STARÝ and SCHLINGER, new species

This species is somewhat related to the European *T. (B.) angelicae* (HAL.) but differs from the latter by the sculpturing of tergite 1 and by the characters of the female genitalia.

Description – Female: Head transverse, smooth, shiny, sparsely haired, wider than thorax. Temple narrower than transverse eye-diameter (7:10). Gena as wide as $1/7$ of longitudinal eye-diameter. Eyes large, widely oval, with sparse short hairs, convergent towards clypeus. Interocular line twice as long as transfacial line (15:7), a little shorter than facial line (15:17). Clypeus with 4 long hairs. Tentorio–ocular line equal to $1/4$ of intertentorial line. Antennae 11-segmented, filiform, reaching half of abdomen. F_1 equal to F_2, more than 3 times as long as wide. Socket-ocular line shorter than socket-diameter (2:3).

Mesoscutum falling vertically to prothorax, smooth, shiny, with sparse long hairs near margins and along effaced notaulices on the disc. Notaulices narrow, crenulate at the ascendent part, effaced on the disc, their fore margin prominent so that central lobe of mesoscutum is edged frontally. Propodeum (Fig. 133) with distinct central areola, carinae prominent. Wing: Pterostigma triangular, more than 3 times as long as wide. Metacarp almost $^1/_2$ shorter than pterostigma. Radial vein longer than pterostigma.

Abdomen lanceolate. Tergite 1 (Fig. 91) 3.5 times as long as wide at spiracles, slender, feebly rugose, with distinct prominent central longitudinal carina. Secondary tubercles distinct, primary tubercles hardly distinct. Distance between primary and secondary tubercles longer than width at spiracles. Genitalia as figured (Fig. 13). Prongs slightly arcuate, with 6 long hairs and one apical hair.

Coloration: Head dark brown. Clypeus, lower part of genae and mouthparts yellow. Antennae brown; scape, pedicel, F_1 and base of F_2 yellow. Thorax dark brown. Wings hyaline, venation brown. Legs yellow. Tergite 1 brown, basal half of abdomen yellow with brown lateral spots, the rest brown. Prongs yellow, base of ovipositor sheaths darkened.

Male: Unknown

Holotype ♀: Japan, Asakawa, May 12, 1961, ex. *Acyrthosiphon* sp., on *Parabenzon praecox* (E. I. SCHLINGER).

Habitat: This species was only encountered once during the Far East Asian expedition, and although several specimens were collected as mummies, only a single ♀ emerged. The habitat was definitely forest-like.

Note: Mummified aphids are light brown.

Trioxys (Binodoxys) carinatus STARÝ and SCHLINGER, new species

This species is closely related to *T. (B.) orientalis* n. sp., but differs from the latter by the shape and structure of tergite 1, the dense and long hair-covered eyes and coloration.

Description – Female: Head transverse, rounded, smooth, shiny, sparsely haired, wider than thorax at tegulae. Occiput feebly margined. Temple $^1/_3$ narrower than transverse eye-diameter. Gena as wide as $^1/_6$ of longitudinal eye-diameter. Eyes large, oval, with dense long hairs. Interocular line nearly twice as long as transfacial line (20:11), $^1/_5$ shorter than facial line.

Facial line more than twice longer than transfacial line. Clypeus with 7 to 8 long hairs. Tentorio-ocular line equal to $^1/_4$ of intertentorial line. Antennae 11-segmented, filiform, reaching apex of tergite 1. F_1 equal to F_2, about 3.5 times as long as wide. Socket-ocular line shorter than half of socket-diameter.

Mesoscutum falling vertically to prothorax, with comparatively dense long hairs, especially near margins and along effaced notaulices on the disc, central lobe edged frontally as seen from above. Notaulices wide, deep, crenulate at the ascendent part, effaced on the disc. Propodeum (Fig. 136) with strongly developed carinae that form more or less complete wide central areola; discs of areolae smooth, with dense long hairs. Wing (Fig. 184): Pterostigma triangular, nearly 3.5 times as long as wide, metacarp somewhat longer than half of pterostigma. Radial vein longer than pterostigma.

Abdomen lanceolate. Tergite 1 (Fig. 93) 2.5 times as long as wide at spiracles, with long and rather prominent central longitudinal carina. Primary and secondary tubercles distinct, the latter rather prominent. Surface smooth, shiny, with sparse long hairs. Following tergites comparatively densely haired. Genitalia as figured (Fig. 17): prongs rather long, with 7 to 8 long hairs on dorsal margin and with 2 apical short hairs.

Coloration: Variable. Upper half of head brown to brownish-yellow. Lower half yellow to yellowish-white, including the mouthparts. Antennae brown, scape, pedicel and base of F_1 yellow. Thorax brown, prothorax yellow, mesoscutum sometimes lighter at base, mesopleurae and propodeum lighter. Wings hyaline, venation light brown. Legs yellow, apices of all tarsi and apices of hind femora brown. Tergite 1 brown except yellow apex, tergite 2 yellow with brown lateral spots, tergite 3 and 4 brown, rest of abdomen yellow; a spot on tergite 9 and apices of prongs a little brown.

Length of body about 2.8 mm.

Male: Unknown.

Holotype ♀: Taiwan, near Tapeito, March 19, 1961, ex. *Macrosiphum rubiformosanum* on *Ribus* sp., (E. I. SCHLINGER). There are also 4 ♀ paratopotypes.

Habitat: This species was only observed once, and that was in a shrub-forest habitat.

Note: Mummified aphids are light brown to brown.

Trioxys (Binodoxys) communis GAHAN

Trioxys communis GAHAN 1926, Proc. U. S. Nat. Mus. 70:4-5 (♀♂, Taiwan, host).

This species belongs to the rather diverse and unclear group, termed the "*T. rietscheli* MACKAUER" group. Its differences are therefore not clear, but it seems to be distinguishable from its relatives by the less distinct primary and secondary tubercles on tergite 1.

Description – Female: Head transverse, rounded, smooth, shiny, sparsely haired, wider than thorax at tegulae. Temple a little narrower than transverse eye-diameter (5: 6). Gena somewhat shorter than $1/4$ of longitudinal eye-diameter (2: 9). Eyes of medium size, prolongately oval, with sparse short hairs, feebly prominent, convergent towards clypeus. Interocular line a little longer than transfacial line (11: 8 to 9), nearly equal to facial line. Clypeus with 4 long hairs. Tentorio-ocular line equal to $1/3$ of intertentorial line. Antennae 11-segmented, filiform, reaching to half of abdomen. F_1 equal to F_2, 3 times as long as wide. Socket-ocular line equal to socket-diameter.

Mesoscutum falling vertically to prothorax, smooth, shiny, nearly hairless. Notaulices narrow, feebly crenulate, hardly visible at the ascendent part and effaced on the disc; their fore margin slightly raised, so that central lobe of mesoscutum only slightly edged frontally. Propodeum (Fig. 132) areolated, discs of areolae smooth, shiny, sparsely haired. Wing (Fig. 185): Pterostigma 2.5 times as long as wide at spiracles. Metacarpus little longer than $1/2$ of pterostigma. Radial vein equal to length of pterostigma or little longer.

Abdomen lanceolate. Tergite 1 (Fig. 97) a little more than twice as long as wide at spiracles, smooth, shiny, slightly convex. Primary tubercles slightly prominent. Secondary tubercles nearly fused with the latter, rather related. Distance between primary and secondary tubercles less than width at spiracles. Genitalia as figured (Fig. 16). Prongs with 4 long hairs at dorsal margin and with 2 apical hairs.

Coloration: Head brown, mouthparts yellowish-brown. Antennae brown; scape, pedicel and F_1 at lower part yellow. Thorax brown. Wings nearly hyaline, venation brown. Legs brown, trochanters and bases of tibiae yellow. Tergite 1 yellow, otherwise abdomen brown, except ovipositor sheaths and prongs somewhat lighter.

Length of body about 1.1 to 1.2 mm.

Male: Antennae 13-segmented, but otherwise similar to female.

General distribution: Taiwan.

Material examined: (201 ♂♀ specimens). Taiwan: Taipei, March 5 to 20, 1961, ex. *Aphis gossypii* on *Hibiscus* sp., 200 ♂♀ (E. I. SCHLINGER); Yung Jean, March 11, 1961, ex. *Rhopalosiphum* sp., on *Verbena phlogiflora*, 1 ♀ (E. I. SCHLINGER).

Type specimen: Taihoku (=Taipei), Formosa, August 30, 1922, bred from *Aphis gossypii*, Cat. No. 28987 USNM.

Habitat: This species was found in two locations, both in ornamental gardens. This was a very common parasite when located, but it should be noted that it was not often found with *Aphis gossypii* in Taiwan.

Trioxys (Binodoxys) indicus SUBBA RAO and SHARMA

Trioxys (Trioxys) indicus SUBBA RAO and SHARMA 1958, Indian J. Ent. 20: 199-202 (♀♂, India-Delhi, host).

The shape of tergite 1 places this species near *T. (B.) struma* GAHAN and *T. (B.) sinensis* MACKAUER from which it differs distinctly by coloration and characters of the female genitalia.

Description – Female: Head transverse, rounded, smooth, shiny, sparsely haired, somewhat wider than thorax at tegulae. Occiput margined. Temple little narrower than transverse eye-diameter. Gena as wide as $1/4$ of longitudinal eye-diameter. Eyes large, widely oval, with sparse short hairs, convergent towards clypeus. Interocular line $1/2$ longer than trans-facial line little, shorter than facial line. Clypeus with 4 long hairs. Tentorio-ocular line equal to $1/4$ of intertentorial line. F_1 equal to F_2 4 times as long as wide. Socket-ocular line shorter than socket-diameter (2:3).

Mesoscutum falling vertically to prothorax, smooth, shiny, with sparse long hairs along margins and effaced notaulices on disc. Notaulices deep, crenulate at ascendent part, effaced on the disc; raised at fore margin so that central lobe of mesoscutum has edged appearance when viewed from above. Propodeum (Fig. 130) distinctly areolated. Wing (Fig. 190): Pterostigma triangular, 3 times as long as wide. Metacarpus shorter than pterostigma. Radial vein equal to length of pterostigma.

Abdomen lanceolate. Tergite 1 (Fig. 102) slender, about 2.5 times as long as wide at spiracles, shiny, with central carina, with distinct secondary tubercles. Distance between primary and secondary tubercles longer than width at spiracles. Width at apex equal to width at spiracles. Following tergites smooth, shiny, sparsely haired. Genitalia as figured (Fig. 14). Prongs nearly straight, with 4 long hairs at dorsal margin and with 2 short apical hairs.

Coloration: Head brown, clypeus and mouthparts yellow. Antennae brown; scape, pedicel and part of F$_1$ yellow. Thorax brown, prothorax and propodeum lighter brown. Wing venation brown. Legs brown, coxae, trochanters, bases of femora and tibiae yellow. Tergite 1 and middle part of tergite 2 yellow, rest of abdomen brown.

Length of body about 1.2 mm.

Male: Antennae 12-segmented. Like female except that yellow coloration is changed to yellowish-brown.

General distribution: India and Taiwan.

Material examined: Taiwan: Taruko, March 24, 1961, ex. *Aphis gossypii*, on *Hibiscus* sp., 1 ♀, 1 ♂, (E. I. SCHLINGER).

Holotype ♀: Delhi, India, bred from *Aphis gossypii* GLOV. on brinjal (=egg plant). Deposited in the National Pusa collections, Div. of Entomology, Indian Agric. Res. Institute, New Delhi.

Allotype male is topotypical.

Habitat: This species may perhaps be rare in Taiwan, since it was only encountered once on a rather commonly observed host *(Aphis gossypii)*. The area near Taruko, Taiwan, where *T. indicus* was found, was on the bank of a river in a rather steppe-forest habitat.

Trioxys (Binodoxys) orientalis STARÝ and SCHLINGER, new species

This species is related to *T. (B.) carinatus* n. sp., but differs from the latter by the sculpturing of tergite 1 and pubescence of eyes and clypeus.

Description – Female: Head transverse, rounded, smooth, shiny, sparsely haired, wider than thorax at tegulae. Occiput feebly margined. Temple $^1/_4$ narrower than transverse eye-diameter. Gena as wide as $^1/_6$ of longitudinal eye-diameter. Eyes large, widely oval, with sparse short hairs, convergent towards clypeus. Interocular line nearly twice as long as transfacial line, shorter than facial line. Facial line more than twice as long as transfacial line. Clypeus with 4 to 5 long hairs. Antennae 11-segmented, filiform, reaching apex of tergite 1. F$_1$ equal to F$_2$, 5 times as long as wide. Socket-ocular line shorter than $^1/_2$ of socket-diameter.

Mesoscutum falling comparatively vertically to prothorax, smooth, shiny, with sparse long hairs, along margins and effaced notaulices on the disc. Central lobe with edged appearance at fore part. Notaulices deep, wide, crenulate at ascendent part, effaced on disc. Propodeum (Fig. 139) with more or less prominent carinae often not forming entirely complete, wide central areola, otherwise smooth, sparsely haired. Wing (Fig. 176): Pterostigma triangular, 3 times as long as wide. Metacarpus longer than half of pterostigma. Radial vein a little longer than pterostigma.

Abdomen lanceolate. Tergite 1 (Fig. 94) about 2.5 to 3 times as long as wide at spiracles, with feeble central longitudinal carina. Primary and secondary tubercles distinct; rugose, shiny on the fore part, with sparse hairs. Following tergites smooth, shiny, sparsely haired. Genitalia as figured (Fig. 21). Prongs long, slightly arcuate, with 6 to 8 long hairs on dorsal margin and with 2 short apical hairs.

Coloration: Variable. Head brown, to nearly yellow with brown markings; face, lower part of temples, mouthparts (except darkened apices of mandibles) yellow. Scape, pedicel and base of F_1 yellow, the rest of antennae brown. Thorax yellowish-orange, usually at least mesoscutum, scutellum and metanotum brown. Wings hyaline, venation brown. Legs yellow, part of hind femora, apices of tibiae and tarsi darkened. Abdomen yellow, lateral spots on tergites 2 to 4 and base of ovipositor sheaths brown. Prongs yellow, sometimes a little darkened at base.

Length of body about 2.1 to 2.6 mm.

Male: Antennae 13-segmented. Head brown, mouthparts and sometimes their close neighborhood yellowish-brown. Antennae brown, scape and pedicel yellow. Thorax brown, lower part more or less lighter colored. Legs yellowish-brown, part of coxae, tibiae and tarsi darkened. Tergite 1 and base of 2 yellowish-brown, remaining tergites brown, otherwise like female except for sexual differences.

General distribution: Japan and South Korea.

Material examined (49 ♂♀ specimens). Japan: Kagoshima, April 13, 1961, ex. *Macrosiphum rosaeibarae* on *Rosa* sp., holotype ♀, allotype ♂, 2 ♀ paratypes (E. I. SCHLINGER). South Korea: Cheju-do Island, May 6, 1961, ex. *Macrosiphum rosaeibarae* on *Rosa* sp., 20 ♂♀ paratypes (E. I. SCHLINGER).

Habitat: In all cases this parasite was collected on ornamental plantings of *Rosa* spp.

Host specificity: Apparently a monophagous parasite of *Macrosiphum rosaeibarae*.

Note: Mummified aphids are quite dark reddish brown.

Trioxys (Binodoxys) sp. – "*rietscheli* MACKAUER", group

Trioxys (Trioxys) rietscheli MACKAUER 1959, Beitr. Ent. 9:170-1 (♀♂ Germany, host). MACKAUER, 1960, Beitr. Ent. 10:141 (subg. *Binodoxys*).

This species group includes members of the genus *Trioxys* which have primary and secondary tubercles on tergite 1 closely connected, and the part of tergite between them nearly parallel-sided.

Description – Female: Head transverse, smooth, shiny, sparsely haired, wider than thorax at tegulae. Temple about $^1/_4$ narrower than transverse eye-diameter. Gena equal to about $^1/_5$ of longitudinal eye-diameter. Eyes large, oval, sparsely haired, convergent towards clypeus. Interocular line about $^1/_3$ longer than transfacial line, little shorter than facial line. Clypeus with 4 long hairs. Tentorio-ocular line equal to $^1/_3$ of intertentorial line. Antennae 11-segmented, filiform, reaching to half of abdomen. F_1 equal to F_2, 2.5 times as long as wide. Socket-ocular line shorter than socket-diameter (2:3).

Mesoscutum smooth, shiny, sparsely haired along effaced notaulices on the disc and along margins. Notaulices narrow, crenulate at the ascendent part, slightly raised at the fore margin, effaced on the disc. Propodeum areolated. Wing: Pterostigma 3 times as long as wide. Metacarpus equal to half of pterostigma length. Radial vein as long as pterostigma.

Abdomen lanceolate. Tergite 1 (Fig. 98) twice as long as wide at spiracles, smooth, shiny, convex. Secondary tubercles distinct. The part of tergite between primary and secondary tubercles nearly parallel-sided. The distance between tubercles shorter than width at spiracles. Genitalia as figured (Fig. 18). Prongs slightly arched, with 4 to 5 long hairs on the dorsal margin, with 1 to 2 hairs at apex.

Coloration: Head dark brown, mouthparts brown. Antennae brown; scape, pedicel, F_1 and base of F_2 yellow. Thorax dark brown. Wings hyaline, venation brown. Legs brown; trochanters and bases of tibiae yellow. Tergite 1 brown, the rest of abdomen dark brown.

Length of body about 1.2 mm.

Male: Antennae 13-segmented, otherwise as described for female.

General distribution: Europe, Far East Asia (USSR-Primorye).

Material examined: USSR – Primorye: Borisovka, 18 km west to Ussurijsk, July 20, 1961, 5 spns., (TRYAPICYN). G. Vjazemskyi, ex. aphids on *Spiraea*, July 20, 1961 (CHUVACHINA).

Hosts: Known originally from *Aphis nasturtii* (KALT.) from Germany (MACKAUER 1959). The host complex of the European material is not cited because of the contemporary state of our knowledge.

Trioxys (Binodoxys) sinensis MACKAUER

Trioxys (Binodoxys) sinensis MACKAUER, 1962, Entomophaga 7(1):40-42.($\female\male$, Hong Kong, W. Pakistan).

This species is related to the European *T. (B.) angelicae* (HAL.), but differs from the latter by the much narrower tergite 1, and by its coloration.

Description – Female: (Translation of MACKAUER) "Head yellow, temples and ocelli somewhat darkened, shiny; narrowed beyond eyes, somewhat wider than thorax; occiput margined. Eyes not prominent, with short hairs. Clypeus, mandibles and palpi yellowish white.

Antennae 13-segmented [segments 1 through 13], reaching to about middle half of abdomen. Segments 1, 2, and anellus light yellow, segment 3 yellowish brown, the following dark brown. Segment 3 about as long as segment 4; apical segment pointed distally and about $1/3$ longer than the praeapical one.

Thorax yellow, shiny, sometimes the sutures somewhat darkened. Mesonotum in the center flat. Praescutellar groove shallow, smooth. Scutellum hardly prominent, margined laterally and apically. Propodeum distinctly areolated, the upper carinae very prominent (Fig. 4). [See our Fig. 140].

Wings yellow hyaline, venation and pterostigma (Index 1:0. 30:0. 51) yellowish brown.

Abdomen light or dark yellow brown, shiny. Tergite 1 yellow brown, with strongly developed primary and secondary tubercles; their relative distance longer than width at spiracles; surface with sparse rugosities and feeble central carina, convex at the apical third, smooth (Fig. 3) [Our figure 100]. Following tergites yellow except the 4th and 5th and more or less large spot at the base of ovipositor sheaths, which are more brown colored. Basal part of ovipositor sheaths quadrangular. Prongs straight, with 5 long hairs at the dorsal margin and 2 apical hairs (Fig. 6 a, b). [Our figure 19].

Legs uniformly yellow and remarkably long haired; the spines of tibiae of 2nd and 3rd pairs somewhat darkened.

Length 2.0 to 2.1 mm; antennae: 1.4 mm.

Male: like the female, except somewhat darker coloration. Head castaneous brown, face and mouthparts yellow. Antennae 13-segmented [segments 4 through 13], reaching to apex of tergite 4, blackish brown. Segment 1 and 2, and lower part of segment 3 yellowish brown. Thorax yellowish brown. Mesonotum, scutellum and fore part of propodeum castaneous brown, the latter with distinct central pentagonal areola. Tergite 1 yellowish brown, with primary and secondary tubercles; the distance between tubercles distinctly longer than width at spiracles. Surface feebly prolongately striated, with feeble central carina. Following tergites darker castaneous brown. Legs yellowish brown, with long hairs; tibiae of 2nd and 3rd pairs darker brown.

Length: 1.6 to 1.8 mm, antennae 1.3-1.5 mm".

General distribution: Hong Kong and West Pakistan.

Material examined: Hong Kong: Taipo, March 7, 1954, ex. *Toxoptera* sp. on *Citrus* sp., 1 ♂ (S. E. FLANDERS).

Holotype ♀: Taipo, Feb. 15, 1954, (S. E. FLANDERS). Deposited in USNM.

Allotype ♂ and paratype ♀ are topotypical.

Note: Some additional notes may be added to the original description from the one male specimen examined by us: Temple narrower than transverse eye-diameter (7:10). Gena equal to $1/4$ of longitudinal eye-diameter. Interocular line less than $1/3$ longer than transfacial line (14:11), somewhat shorter than facial line. Tentorio-ocular line somewhat shorter than $1/2$ of intertentorial line (2:5). Socket-ocular line shorter than socket-diameter (2:3).

Trioxys (Binodoxys) struma GAHAN

Trioxys struma GAHAN, 1926, Proc. U. S. Nat. Mus. 70:5-6 (♀♂, Taiwan, host).

This species belongs to the group of *Binodoxys* species which is characterized by having the distance between the primary and secondary tubercles longer than the width at the spiracles. It is similar to *T. (B.) sinensis* MACKAUER, but differs from the latter chiefly by having 2 remarkably long apical hairs on the prongs.

Description – Female: Head transverse, rounded, smooth, shiny, sparsely haired, wider than thorax at tegulae. Occiput feebly margined. Temple $1/5$ narrower than transverse eye-diameter. Gena as long as $1/4$ of longitudinal eye-diameter. Eyes of medium size, oval, with sparse short hairs, convergent towards clypeus. Interocular line $\frac{1}{2}$ longer than transfacial line, little shorter than facial line. Clypeus with 4 to 6 long hairs. Tentorio-ocular line somewhat less than $\frac{1}{3}$ of intertentorial line. Antennae 11-segmented, filiform. F_1 equal to F_2, 3 times as long as wide. Socket-ocular line as long as half of socket-diameter.

Mesoscutum falling arcuately-vertically to prothorax, smooth, shiny, with sparse and remarkably long hairs along margins and effaced notaulices on the disc, central lobe edged frontally. Notaulices distinct at the ascendent part, deep, crenulate, effaced on the disc. Propodeum (Fig. 135) areolated, with wide central areola; disc of areolae shiny, sparsely haired. Wing (Fig. 189): Pterostigma triangular, 3.5 times as long as wide. Metacarpus somewhat longer than half of pterostigma length. Radial vein longer than pterostigma.

Abdomen lanceolate. Tergite 1 (Fig. 103) slender, more than 3 times as long as wide at spiracles, with long prominent central carina and secondary tubercles, feebly rugose-granulate, shiny, sparsely haired. Genitalia as figured (Fig. 50). Prongs rather long, slightly curved upwards at apex, with 5 long hairs on the dorsal margin and 2 rather long apical hairs.

Coloration: Head yellowish-orange. Scape, pedicel, F_1 and F_2 yellow, the rest of antennae brown. Thorax yellowish-orange. Wings hyaline. Venation yellow. Legs yellow, apices of tarsi darkened, hind femora and tibiae brown. Abdomen yellowish-orange, tergites 3 and 4 and ovipositor sheaths partly brown. Prongs yellowish-orange.

Length of body about 2.1 mm.

Male: "Antennae 13-segmented, as long as the body or nearly. Head and thorax above brownish yellow, beneath paler; scape and pedicel yellow, the rest of antennae black; abdomen mostly brownish black, the first and large part of second tergites testaceous; legs concolorous with underside of thorax, other characters as in the female." (from GAHAN, 1926).

General distribution: Taiwan.

Material examined: Taiwan: Qua Ing Shan, nr. Taipei, March 26, 1961, ex. *Megoura citricola* on *Ficus* (?) sp., 2 ♀♀, (E. I. SCHLINGER).

Type specimen: "Taihoku, Formosa, Cat. No. 28986 USNM, bred from *Macrosiphoniella citricola* v.d.G." [= *Megoura citricola*].

Habitat: The only time this species was located was in a dense forest. Associated with the host aphid, *Megoura citricola*, were many *Greenidea ficicola*, but none of the latter were parasitized. *G. ficicola* was recorded on a host of this species by GAHAN (1926), but because of the above association, other similar rearing trials without rearing any parasites, and the fact that there are several distinct species now known to attack *G. ficicola*, we tend to doubt that *G. ficicola* is a host of this parasite.

Note : Mummified *Megoura citricola* aphids are brownish-black.

Trioxys (Binodoxys) sp.?

This species appears to be distinct from other Taiwan *T. (B.)* species males, but cannot be named in the absence of female specimens: a description of the male, however, may help place this species at a later date.

Male : Head transverse, smooth, shiny, sparsely haired, wider than thorax at tegulae. Temple equal to transverse eye-diameter. Gena as long as $^1/_5$ of longitudinal eye-diameter. Eyes of medium size, widely oval, with sparse short hairs, convergent towards clypeus. Interocular line about $^1/_3$ longer than transfacial line (11:8), somewhat shorter than facial line (11: 13). Clypeus with 3 long hairs. Tentorio-ocular line equal to $^1/_3$ of inter-tentorial line. Antennae 13-segmented, filiform, reaching to $^2/_3$ of abdomen. F_1 equal to F_2, 3 times as long as wide. Socket-ocular line shorter than socket-diameter (2:3).

Mesoscutum falling vertically to prothorax, smooth, shiny, sparsely haired, with sparse long hairs along margins and effaced notaulices on the disc. Notaulices narrow, crenulate at the ascendent part, prominent at the fore margin so that the central lobe of mesoscutum has edged appearance as seen from above; effaced on the disc. Propodeum (Fig. 131) areolated. Wing: Pterostigma nearly 3 times as long as wide. Metacarpus about $^1/_3$ shorter than pterostigma. Radial vein longer than pterostigma.

Abdomen rounded at apex. Tergite 1 (Fig. 92) 2.5 times as long as wide at spiracles, with feeble central longitudinal carina, slightly rugose, shiny, sparsely haired, with distinct primary and secondary tubercles. Distance between primary and secondary tubercles longer than width at spiracles.

Coloration : Head brown. Mouth parts and neighborhood yellow. Antennae brown, lower part of scape and pedicel yellow. Thorax brown, prothorax, metapleurae and propodeum lighter colored. Wings hyaline, venation brown. Legs yellow, middle and hind femora and tibiae brown. Tergite 1 and central base of tergite 2 yellow, the rest brown.

Female : Unknown.

Material examined. Taiwan: Wushe, (= Musha), March 14, 1961, ex. *Aphis spiraecola* on *Spiraea* sp., 1 ♂, (E. I. SCHLINGER).

Habitat: The unique specimen was collected in a shrub-forest area.

SUBGENUS *Fissicaudus* STARÝ and SCHLINGER, new subgenus

This subgenus is related to *Binodoxys* MACKAUER, but differs from the latter primarily by the characters of the female prongs.

Description: Head transverse. Antennae filiform, with variable number of segments (11 to 13). Eyes large. Mesoscutum with edged central lobe at the fore part as seen from above. Notaulices distinct at the ascendent part. Propodeum areolated. Wing: Pterostigma triangular, longer than meta-carpus. Radial vein distinctly developed. Otherwise venation effaced towards wing-apex except cubital cell 2. Abdomen of female lanceolate. Tergite 1 with primary and secondary tubercles. Ovipositor sheaths bean-shaped, curved downwards. Prongs initiate near the lateral base of last sternite.

Type species: Trioxys (Binodoxys) confucius MACKAUER, by present designation.

General distribution: Hong Kong and Taiwan.

Bionomics: Parasite of a rather specialized group of aphids of the genus *Greenidea*. Pupation occurs inside the parasitized aphid.

Trioxys (Fissicaudus) confucius MACKAUER

Trioxys (Binodoxys) confucius MACKAUER, 1962, Entomophaga 7 (1): 37-39 (♀ Hong Kong, host).

Description – Female: Head transverse, rounded, smooth, shiny, sparsely haired, somewhat wider than thorax at tegulae. Occiput margined. Temple $^1/_3$ narrower than transverse eye-diameter. Gena equal to $^1/_5$ of longitudinal eye-diameter. Eyes large, nearly hemispherical, prominent, with sparse short hairs, convergent towards clypeus. Interocular line $^1/_3$ longer than transfacial line shorter than facial line. Clypeus with 6 long hairs. Tentorio-ocular line somewhat shorter than half of intertentorial line. Antennae 11-segmented, filiform, reaching to half of abdomen. F_1 equal to F_2, 3 times as long as wide. Socket-ocular line longer than half of socket-diameter.

Mesoscutum falling vertically to prothorax, smooth, shiny, sparsely haired along margins and effaced notaulices on the disc. Notaulices wide, crenulate at the ascendent part, with prominent fore margin so that the central lobe of mesoscutum has edged appearance as seen from above; effaced on the disc. Scutellum widely prolongately triangular, somewhat prominent, smooth, shiny, sparsely haired. Propodeum (Fig. 138) with rather prominent carinae, with complete large central areola. Wing (Fig. 188): Pterostigma 3 to 3.5 times as long as wide. Metacarp about $^1/_3$ shorter than pterostigma. Radial vein longer than pterostigma.

Abdomen lanceolate. Tergite 1 (Fig. 96) slender, more than 3 times as long as wide at spiracles, nearly smooth, with central longitudinal carina, sparsely haired. Secondary tubercles distinct. Distance between primary and secondary tubercles longer than width at spiracles. Following tergites smooth, shiny, with sparse long hairs. Genitalia as figured (Fig. 15). Ovipositor sheaths bean-shaped. Prongs initiating near the lateral base of last sternite.

Coloration: Head brown, face, lower part of temples, genae and mouthparts yellow. Antennae brown, scape, pedicel, F_1 and part of F_2 yellow. Thorax yellow. Wings hyaline, venation brownish-yellow. Legs light yellow, apices of tarsi darkened. Tergite 1 and 2 yellow, the rest of abdomen dark brown, except yellow prongs.

Length of body about 2 mm.

Male: Antennae 13-segmented. Coloration of head and thorax darker than in female, otherwise as described for female.

General distribution: Hong Kong and Taiwan.

Material examined (45 ♂♀ specimens). Hong Kong: Kowloon, Feb. 25, 1961, ex. *Greenidea ficicola* on *Ficus* sp., 15 ♂♀, (E. I. SCHLINGER). Taiwan: Yan Ming Shan, March 8, 1961, ex. *Greenidea ficicola*, 1 spn., (E. I. SCHLINGER), Taipei, March 27, 1961, ex. *Greenidea ficicola* on *Ficus* sp., 12 ♂♀ (E. I. SCHLINGER).

Holotype ♀: Taipo, July 18, 1954, ex. aphid on *Citrus*, (S. FLANDERS) (No. 588)2). Deposited in the USNM.

Habitat: This species was not uncommon in Taipei in forest-like habitats, but we doubt the record of the holotype coming from an "Aphid on *Citrus*" and feel this rearing should be carefully checked.

Host specificity: Judging from the morphological adaptations and habitat associations, this species seems to be a monophagous parasite of *Greenidea ficicola*.

Note: Mummified aphids are light brown.

SUBGENUS *Trioxys* HALIDAY

Description: Head transverse. Antennae filiform, (♀) 11 to 12- (♂) 12 to 14-segmented. Propodeum smooth or areolated. Tergite 1 with primary tubercles located on anterior third of segment.

General distribution: Holarctic and Oriental regions.

Trioxys (Trioxys) asiaticus TELENGA

Trioxys asiaticus TELENGA, 1953, Trudy Inst. Zool. i parazitol. AN Uzb SSR 1: 170-1 (♀, USSR – Uzbekistan). LUZHETZKI, 1959, Tez. dokl. 4-ogo sj. Vses. ent. Obststh., Moscow-Leningrad, p. 82. LUZHETZKI, 1960, Par. tlej Uzbekistana, ,pp. 140-1. (♀♂, USSR – Uzbekistan, hosts).

Trioxys (Trioxys) vandenboschi MACKAUER, 1960, Senck. biol., Frankfurt M., 41:359-61 (♀♂, Iran, host).

This species is characterized by its wing-venation. It differs easily from a Far East Asian relative, *T. (T.) luteolus* n. sp., by the characters on female genitalia and prongs. Otherwise it is related to the European *T. (T.) pannonicus* STARÝ, but differs from the latter by the 12-segmented antennae and the host-complex. Its relationship to the Nearctic species (i.e. *T. infrequens* SMITH) has not been studied intensively yet by the authors.

Description – Female: Head nearly subsquare as seen from above, smooth, shiny, sparsely haired, somewhat wider than thorax at tegulae. Occiput margined. Temple $1/5$ narrower than transverse eye-diameter. Gena as wide as $1/5$ of longitudinal eye-diameter. Eyes large, widely oval, strongly prominent especially at the lower part, convergent towards clypeus. Interocular line almost twice as long as transfacial line, little shorter than facial line. Clypeus transverse, smooth, shiny, nearly flat, with about 6 long hairs. Tentorio-ocular line equal to $1/6$ of intertentorial line. Antennae 12-segmented, filiform, reaching to middle of abdomen, situated at the level of eye-center. F_1 rather slender, about 6 times as long as wide, F_2 a little shorter. Socket-ocular line shorter than socket-diameter.

Mesoscutum raised above prothorax, without covering it when viewed laterally. Notaulices deep, crenulate at the fore part, effaced on the disc. Propodeum (Fig. 137) smooth, shiny, sparsely haired, with 2 short, divergent, feeble carinae at the lower part. Wing (Fig. 187): Pterostigma triangular.

Metacarp rather short, nearly pointlike. Radial vein somewhat longer than half of pterostigma.

Abdomen lanceolate. Tergite 1 (Fig. 99) slender, nearly parallel-sided, about twice as long as wide at spiracles, shiny, slightly convex, sparsely haired. Spiracular tubercles feebly prominent, situated at the end of the first third of tergite. Genitalia as figured (Fig. 12). Prongs long, slender, upwardly curved at apex, with 8 to 9 long hairs on the dorsal margin and 2 lanceolate hairs at apex.

Coloration: Variable. Head yellowish-brown. Antennae brown, scape, pedicel, F_1 and part of F_2 yellowish-brown. Thorax yellowish-brown, with more or less distributed brown areas on mesoscutum and scutellum. Wing-venation brown. Legs yellow, upper part of middle and hind femora sometimes brown. Tergites 1 and 2 yellowish-brown, 3, 4 and 5 brown, the rest of abdomen yellowish-brown (including prongs).

Length of body about 2.1 to 2.6 mm.

Male: Antennae 14-segmented. Dark brown; mouthparts, pedicel, basal part of tergite 1 and more or less the center of abdomen and part of legs yellowish-brown to brown. Otherwise like the female.

General distribution: Asia Minor (Iran), Central Asia (USSR – Uzbekistan), Far East Asia (USSR – Primorye).

Material examined (6 specimens). USSR – Primorye: Borisovka, 18 km west of Ussurijsk, July 20, 1961, 5 spns., (TRYAPICYN); Suputinsky zapovednik, August 8, 1961, mixed forest, 1 ♀ (KOVALEV).

Type specimens:

Trioxys asiaticus TELENGA: ♀, Uzbekistan – Tashkent (PLOTNIKOV). Probably deposited in the collection of Prof. TELENGA at Kiev.

Trioxys vandenboschi MACKAUER. Holotype ♀: Hesarak, June, 1960, (R. VAN DEN BOSCH). Deposited in the collection of Dr. MACKAUER (Frankfurt, Germany). Allotype ♂ is topotypical.

Habitat: Known from Central Asia as a parasite of cotton aphids occurring in field habitats.

Hosts: (Unrevised literary data).

Acyrthosiphon gossypii MORDV.: MACKAUER, 1960, on *Sophora alopecuroides*, Iran.

Aphis medicaginis KOCH: LUZHETZKI, 1960, on *Gossypium hirsutum*, *Robinia pseudoacacia* and *Sophora*. USSR – Uzbekistan.

Hosts: (Original and revised literary data).

Acyrthosiphon gossypii gossypii MORDV.: LUZHETZKI, 1960, on *Gossypium hirsutum* USSR – Uzbekistan.

Note: When revising the Central Asiatic species we found that *Trioxys asiaticus* TELENGA 1953 is identical with *Trioxys vandenboschi* MACKAUER 1960, so that for priority reasons it is necessary to synonymize the latter name.

Trioxys (Trioxys) luteolus STARÝ & SCHLINGER, new species

This species is related to the European *T. (T.) cirsii* (CURTIS), but differs from the latter by coloration, sculpture of propodeum, shape of tergite 1 and by the host-complex.

Description – Female: Head transverse, rounded, smooth, shiny, sparsely haired, wider than thorax at tegulae. Occiput feebly margined. Temple $^1/_3$ narrower than transverse eye-diameter. Gena as wide as $^1/_7$ of longitudinal eye-diameter. Eyes large, widely oval, with sparse short hairs, convergent towards clypeus. Interocular line about $^1/_3$ longer than transfacial line, little shorter than facial line. Clypeus with about 9 long hairs. Tentorio-ocular line equal to $^1/_6$ to $^1/_7$ of intertentorial line. Antennae 11-segmented, filiform, reaching to middle of abdomen. F_1 equal to F_2, more than 4 times as long as half of socket-diameter or a little shorter.

Mesoscutum a little elevated above prothorax, smooth, shiny, its central lobe edged frontally (as seen from above), with sparse long hairs along effaced notaulices on the disc and along margins. Notaulices distinct at the ascendent part, crenulate, effaced on the disc. Propodeum (Fig. 134) areolated, with irregular carinae that form a more or less complete wide, central areola; discs of areolae smooth, with sparse long hairs. Wing (Fig. 186): Pterostigma triangular, 2.5 to 3 times as long as wide. Metacarp about half as long as pterostigma. Radial vein about as long as pterostigma.

Abdomen lanceolate. Tergite 1 (Fig. 90) 2.5 times as long as wide at spiracles, nearly parallel-sided, smooth, with sparse hairs. Spiracular tubercles hardly visible, situated before the middle point of tergite. Following tergites smooth, shiny, sparsely haired. Genitalia as figured (Fig. 20); prongs very slender and long, slightly curved upwards, with short hairs and 2 apical lanceolate hairs.

Coloration: Head brown, lower part of temples, face, clypeus and mouth-parts yellow. Antennae brown, scape, pedicel and F_1 yellow. Thorax yellowish-orange. Wings hyaline, venation light brown. Legs yellowish orange, apices of tarsi darkened. Tergite 1 yellow, tergite 2 and part of tergite 3 yellow, brown along margins. Tergites 3 to 5 brown, remaining ones yellow. Ovipositor sheaths and prongs brownish-yellow.

Male: Antennae 13-segmented. Head brown, mouthparts brownish-yellow. Antennae brown, lighter at base. Thorax brown, prothorax and part of propodeum lighter. Legs yellowish-brown, apices of tarsi darkened. Tergite 1 and central longitudinal spot on tergite 2 yellowish-brown; the rest of abdomen brown. Otherwise like the female except for sexual differences.

General distribution: Taiwan.

Holotype ♀: Taipei, Taiwan, March 24, 1961, ex. *Shivaphis* n. sp. on *Acer* sp. (E. I. SCHLINGER). Allotype ♂ is topotypical.

Other specimens examined were 83 paratypes as follows: Taipei, Taiwan, 65 ♂♀ paratopotypes. Taipei, Taiwan, March 20, 1961, ex. *Shivaphis* n. sp. on *Keteleeria davidiana*, 2 ♂ paratypes (E. I. SCHLINGER). Taipei, Taiwan, March 20, 1961, ex *Agrioaphis viridis* on *Ulmus* sp. 16 ♂♀ paratypes (E. I. SCHLINGER).

Habitat: All of the above specimens were collected in the Taipei Botanical Garden and were quite arboreal in habit. Even though the host aphids were often observed on lower tree branches, the parasitized aphids were invariably found on leaves over six feet high.

Host specificity: Although three host aphids are indicated above, only *Shivaphis* n. sp. on *Acer* sp. acted as a "good" host for this parasite. Considerably more time was spent looking for parasitized aphids *(Agrioaphis)* on *Ulmus* and *(Shivaphis)* on *Keteleeria* and with higher aphid densities and yet fewer parasites were located. Although this parasite is oligophagous, a definite host preference was observed with the type and topotype specimens.

Note: The mummified aphids were brown, but parasitized *Shivaphis* mummies appear white due to the aphids' white flocculence.

Trioxys (Trioxys) species

This species is probably new, but inadequate specimens prevents us from describing it at this time.

Male: Head transverse, rounded, smooth, shiny, sparsely haired, wider than thorax at tegulae. Temple $^1/_3$ narrower than transverse eye-diameter. Gena as wide as $^1/_6$ of longitudinal eye-diameter. Eyes large, widely oval, prominent, with sparse short hairs, convergent towards clypeus. Interocular line $^1/_3$ longer than transfacial line somewhat shorter than facial line.

Clypeus with 6 long hairs. Tentorio–ocular line equal to $^1/_3$ of intertentorial line. Antennae with more than 9 segments (broken), filiform. F_1 3 times as long as wide, equal to F_2. Socket–ocular line shorter than socket-diameter (as 2 : 3).

Mesoscutum smooth, shiny, vertically falling to prothorax, with sparse long hairs along effaced notaulices on the disc and margins. Notaulices narrow, feebly crenulate at the ascendent part, effaced on the disc. Propodeum (Fig. 129) areolated. Wing: Pterostigma more than 3 times as long as wide. Metacarp half as long as pterostigma. Radial vein somewhat longer than pterostigma.

Abdomen rounded at apex. Tergite 1 (Fig. 95) twice as long as wide at spiracles, slender, nearly parallel-sided, smooth, shiny, sparsely haired. Spiracular tubercles situated at the end of first third of tergite, slightly prominent. Following tergites comparatively densely haired.

Coloration: Head dark brown, mouthparts and neighborhood yellow. Antennae brown, scape, pedicel and base of F_1 yellow. Thorax brown; prothorax and metapleurae lighter brown. Wings hyaline, venation brown. Legs yellow. Tergite 1 and base of tergite 2 yellow, the rest of abdomen brown.

Length of body about 1.4 mm.

Female: Unknown.

Material examined (one specimen). Taiwan: Sun Moon Lake, March 14, 1961, ex. *Phyllaphoides bambusicola* on *Bambusa* sp., 1 ♂, (*E. I. Schlinger*).

Habitat: This specimen (along with several others which yielded only secondary parasites) was collected in a dense bamboo covered hillside near the north shore of Sun Moon Lake. The host aphid was extremely rare and the small populations that were present were heavily parasitized by an *Aphelinus* species.

FAR EAST ASIAN APHID HOST-PARASITE LIST

1. *Acyrthosiphon* spp.
   Aphidius sp. no. 2
   Aphidius sp. no. 5
   Aphidius sp. no. 7
   Praon orientale STARÝ AND SCHLINGER
   Trioxys (Binodoxys) brunnescens STARÝ AND SCHLINGER
2. *Agrioaphis viridis* TAKAHASHI
   Ephedrus (Ephedrus) persicae FROGGATT
   Trioxys (Trioxys) luteolus STARÝ AND SCHLINGER
3. *Amphicercus japonicus* (HORI)
   Ephedrus (Ephedrus) plagiator (NEES)
4. *Amphorophora lactucae* (L.)
   Ephedrus sp. (emer.)
5. *Amphorophora lespedezae* ESSIG AND KUWANA
   Aphidius sp. (emer.)
6. *Amphorophora lonicericola* TAKAHASHI
   Ephedrus (Ephedrus) plagiator (NEES)
7. *Amphorophora magnoliae* ESSIG AND KUWANA
   Ephedrus (Ephedrus) plagiator (NEES)
   Praon orientale STARÝ AND SCHLINGER
8. *Amphorophora oleracea* VAN DER GOOT
   Ephedrus (Ephedrus) orientalis STARÝ AND SCHLINGER
9. *Amphorophora* spp.
   Aphidius sp. no. 6
   Praon orientale STARÝ AND SCHLINGER
10. *Anuraphis mumei* (HORI)
    Ephedrus (Ephedrus) plagiator (NEES)
11. *Aphis gossypii* GLOVER
    Ephedrus (Ephedrus) persicae FROGGATT
    Ephedrus (Ephedrus) plagiator (NEES)
    Lipolexis gracilis FÖRSTER
    Lysiphlebia japonica (ASHMEAD)
    Trioxys (Binodoxys) communis GAHAN
    Trioxys (Binodoxys) indicus SUBBA RAO AND SHARMA
12. *Aphis laburni* KALTENBACH (= *A. cytisorum*)
    Ephedrus (Ephedrus) plagiator (NEES)
13. *Aphis malvoides* VAN DER GOOT
    Lipolexis scutellaris MACKAUER
14. *Aphis nerii* B. DE F.
    (Unknown genus)
15. *Aphis pomi* DE GEER
    Ephedrus (Ephedrus) plagiator (NEES)
16. *Aphis rumicis* LINNAEUS
    Ephedrus (Ephedrus) plagiator (NEES)
17. *Aphis sacchari* ZEHNTER (i.e. *Longiunguis*)
    Lysiphlebia japonica (ASHMEAD)
18. *Aphis sambuci* L.
    Lysiphlebia japonica (ASHMEAD)
19. *Aphis saliceti* KALTENBACH
    Praon sp. (emerged)
20. *Aphis spiraecola* PATCH
    Ephedrus (Ephedrus) persicae FROGGATT
    Ephedrus (Ephedrus) plagiator (NEES)
    Lipolexis gracilis FÖRSTER
    Lysiphlebia japonica (ASHMEAD)
    Praon orientale STARÝ AND SCHLINGER
    Trioxys (Binodoxys) sp.
21. *Aphis* spp.
    Lysiphlebia japonica (ASHMEAD)
    Lysiphlebia rugosa STARÝ AND SCHLINGER
    Lysiphlebus sp. aff. *delhiensis* SHARMA AND SUBBA RAO
    Praon sp. no. 1
22. *Brachycaudus helichrysi* (KALTENBACH)
    Lysiphlebia japonica (ASHMEAD)
    Lysiphlebia rugosa STARÝ AND SCHLINGER
23. *Brachysiphoniella graminis* TAKAHASHI
    Ephedrus (Ephedrus) plagiator (NEES)
24. *Brevicoryne brassicae* (LINNAEUS)
    Diaeretiella rapae (M'INTOSH)
25. *Capitophorus* spp.
    Aphidius sp. "*gifuensis* ASHMEAD" group
    Ephedrus (Ephedrus) persicae FROGGATT
    Ephedrus (Ephedrus) plagiator (NEES)
26. *Cavariella araliae* TAKAHASHI

Aphidius salicis HALIDAY
Ephedrus (Ephedrus) plagiator (NEES)

27. *Cavariella bicaudata* ESSIG AND KUWANA
Aphidius sp. (prob. *salicis*)

28. *Cavariella capreae* (FABR.)
Aphidius sp. (prob. *salicis*)

29. *Cavariella salicicola* (MATSUMURA)
Aphidius salicis HALIDAY

30. *Cavariella* spp.
Aphidius salicis HALIDAY
Praon orientale STARÝ AND SCHLINGER

31. *Chaitophorus salicicolus* MATSUMURA
Lysiphlebus salicaphis FITCH

32. *Cinara formosana* TAKAHASHI
Pauesia unilachni (GAHAN)

33. *Cinara laricicolus* (MATSUMURA)
Pauesia laricis (HALIDAY)
Pauesia pini (HALIDAY)

34. *Cinara laricis* (WALKER)
Pauesia pini (HALIDAY)

35. *Cinara longipennis* (MATSUMURA)
Pauesia konoi (WATANABE)

36. *Cinara nopporoensis* INOUYE
Pauesia jezoensis (WATANABE)

37. *Cinara orientalis* TAKAHASHI
Pauesia unilachni (GAHAN)

38. *Cinara pineti* (KOCH)
Pauesia pini (HALIDAY)

39. *Cinara thujafoliae* THEOBALD
Pauesia sp. (emer.)

40. *Cinara todocolus* (INOUYE)
Pauesia inouyei (WATANABE)

41. *Cinara* spp.
Pauesia infulata (HALIDAY)
Pauesia unilachni (GAHAN)

42. *Coloradoa rufomaculata* WILSON
Praon quadratum STARÝ AND SCHLINGER

43. *Dilachnus* sp.
Pauesia laticeps (GAHAN)

44. *Eriosomatinae* sp.
Ephedrus (Lysephedrus) validus (HALIDAY)

45. *Euceraphis betulae* LINNAEUS
Praon glabrum STARÝ AND SCHLINGER

46. *Eulachnus piniformosanus* TAKAHASHI
Diaeretus leucopterus (HALIDAY)

47. *Eulachnus* sp.
Pauesia unilachni (GAHAN)

48. *Greenidea ficicola* TAKAHASHI
Archaphidus greenideae STARÝ AND SCHLINGER
Trioxys (Fissicaudus) confucius MACKAUER

49. *Hyperomyzus lactucae* (L.)
Ephedrus (Ephedrus) plagiator (NEES)
Praon orientale STARÝ AND SCHLINGER

50. *Hyalopterus pruni* (GEOFFR.)
Ephedrus (Ephedrus) plagiator (NEES)

51. *Lachniella costata* (ZETTERSTEDT)
Pauesia jezoensis (WATANABE)

52. *Lachnus tropicalis* V. D. GOOT
Pauesia japonica (ASHMEAD)
Pauesia tropicalis STARÝ AND SCHLINGER

53. *Macrosiphoniella formosartemisiae* TAKAHASHI
Aphidius absinthii MARSHALL

54. *Macrosiphoniella sanborni* (GILLETTE)
Aphidius absinthii MARSHALL
Ephedrus (Ephedrus) campestris STARÝ

55. *Macrosiphoniella tanacetaria* KOCH
Aphidius absinthii MARSHALL

56. *Macrosiphoniella yomogifoliae* TAKAHASHI
Aphidius absinthii MARSHALL

57. *Macrosiphoniella yomogifoliae* TAKAHASHI
Ephedrus (Ephedrus) campestris STARÝ

58. *Macrosiphum akebiae* SHINJI
Praon sp. (emer.)

59. *Macrosiphum formosanum* TAKAHASHI
Ephedrus (Ephedrus) plagiator (NEES)

60. *Macrosiphum gobonis* (MATS.)
Ephedrus sp. (emer.)

61. *Macrosiphum lactucicola* (STRAUD)
Aphidius sp. (emer.)

62. *Macrosiphum rosae* (L.)
Aphidius sp. no. 3
Ephedrus (Ephedrus) plagiator (NEES)

63. *Macrosiphum rosaeibarae* MATSUMURA
Aphidius sp. "*gifuensis* ASHMEAD" group
Aphidius sp. no.1
Ephedrus (Ephedrus) plagiator (NEES)
Praon orientale STARÝ AND SCHLINGER
Trioxys (Binodoxys) orientalis STARÝ AND SCHLINGER

64. *Macrosiphum rubiformosanum* TAKA-HASHI
    Trioxys (Binodoxys) carinatus STARÝ AND SCHLINGER
65. *Macrosiphum* spp.
    Aphidius absinthii MARSHALL
    Aphidius sp. "*gifuensis* ASHMEAD" group
    Ephedrus (Ephedrus) plagiator (NEES)
66. *Matsumuraja rubicola* TAKAHASHI
    Aphidius sp. (emer.)
67. *Megoura citricola* (V.D. GOOT)
    Ephedrus (Ephedrus) plagiator (NEES)
    Trioxys (Binodoxys) struma GAHAN
68. *Megoura viciae* BUCKTON
    Ephedrus (Ephedrus) campestris STARÝ
69. *Myzocallis yokohamai* TAKAHASHI
    Trioxys sp. (ovipositing)
70. *Myzus malsuctus* MATSUMURA
    Ephedrus (Ephedrus) persicae FROGGATT
71. *Myzus momonis* (MATSUMURA)
    Ephedrus (Ephedrus) persicae FROGGATT
72. *Myzus mumecola* MATSUMURA
    Ephedrus (Ephedrus) persicae FROGGATT
73. *Myzus persicae* (SULZER)
    Aphidius sp. "*gifuensis* ASHMEAD" group
    Ephedrus (Ephedrus) persicae FROGGATT
    Ephedrus (Ephedrus) plagiator (NEES)
74. *Myzus polypodicola*
    Monoctonus sp. (emer.)
    Praon sp. (emer.)
75. *Myzus woodwardiae* TAKAHASHI
    Ephedrus (Ephedrus) lacertosus (HALI-DAY)
    Monoctonus (Monoctonus) woodwardiae STARÝ AND SCHLINGER
76. *Myzus* spp.
    Aphidius sp. "*gifuensis* ASHMEAD" group
    Ephedrus (Ephedrus) persicae FROGGATT
    Monoctonus (Monoctonus) similis STARÝ AND SCHLINGER
    Praon orientale STARÝ AND SCHLINGER
77. *Parachaitophorus spiraeae* TAKAHASHI
    Ephedrus (Ephedrus) plagiator (NEES)
    Lysiphlebia japonica (ASHMEAD)
78. *Pentalonia nigronervosa* COQ.

Unknown emer. mummies
79. *Pergandeida* sp. (i.e. *Aphis* sp.)
    Unknown emer. mummy
80. *Pergandeida trirhodus* WALKER (i.e. *Longicaudus*)
    Praon sp. (emer.)
81. *Periphyllus koelreuteriae* TAKAHASHI
    Aphidius areolatus ASHMEAD
82. *Periphyllus testudinatus* (FERNIE)
    (=*testudinatus* THORNTON)
    Aphidius areolatus ASHMEAD
83. *Periphyllus* sp.
    Praon sp. no. 3
84. *Phyllaphidinae* sp.
    Bioxys japonicus STARÝ AND SCHLINGER
85. *Phyllaphoides bambusicola* TAKAHASHI
    Trioxys (Trioxys) sp.
86. *Prociphilus konoi* HORI.
    Ephedrus (Ephedrus) plagiator (NEES)
87. *Rhopalosiphoninus deutzifoliae* SHINJI
    Aphidius sp. no. 4
    Ephedrus (Ephedrus) plagiator (NEES)
    Praon orientale STARÝ AND SCHLINGER
88. *Rhopalosiphoninus tiliae* MATSUMURA
    Praon orientale STARÝ AND SCHLINGER
89. *Rhopalosiphum* sp.
    Trioxys (Binodoxys) communis GAHAN
90. *Rhopalosiphum prunifoliae* FITCH
    (Unknown sp. whose mummies were egg-shaped and not attached to plant surface)
91. *Shivaphis* sp.
    Trioxys (Trioxys) luteolus STARÝ AND SCHLINGER
92. *Sitobium granarium* (KIRBY)
    Ephedrus (Ephedrus) plagiator (NEES)
93. *Stomaphis yanonis* TAKAHASHI
    Protaphidius nawaii (ASHMEAD)
94. *Tetraneura* sp.
    Lysiphlebia japonica (ASHMEAD)
95. *Toxoptera aurantii* (B. DE F.)
    Lipolexis gracilis FÖRSTER
96. *Toxoptera odinae* V. D. GOOT
    Ephedrus (Ephedrus) plagiator (NEES)
    Lysiphlebia japonica (ASHMEAD)
97. *Toxoptera* spp.
    Lysiphlebia japonica (ASHMEAD)
    Trioxys (Binodoxys) sinensis MACKAUER

98. *Tuberculoides querciformosanus* TAKA-
HASHI
   Trioxys sp. (ovipositing)
99. *Tuberolachnus salignus* (GMELIN)
   Aphidius salignae WATANABE

100. *Unilachnus* sp.
   Pauesia unilachni (GAHAN)
101. *Vesiculaphis caricis* FULLAWAY
   Ephedrus (Ephedrus) persicae FROG-
GATT

TAIWAN (FORMOSA)

Archaphidus greenideae STARÝ AND SCHLINGER
  Greenidea ficicola TAKAHASHI
Diaeretiella rapae (MCINTOSH)
  Brevicoryne brassicae (L.) [TAKAHASHI, 1925]
Lysiphlebia japonica (ASHMEAD)
  Aphis sacchari ZEHNTER (i.e. *Longiunguis*)
Monoctonus woodwardiae STARÝ AND SCHLINGER
  Myzus woodwardiae TAKAHASHI
Ephedrus (E.) lacertosus (HALIDAY)
  Myzus woodwardiae TAKAHASHI
Ephedrus (E.) orientalis STARÝ AND SCHLINGER
  Amphorophora oleracea V.D. GOOT
Ephedrus (E.) persicae FROGGATT
  Agrioaphis viridis TAKAHASHI
  Aphis gossypii GLOVER
  Aphis spiraecola PATCH
  Myzus momonis (MATSUMURA)
  Myzus persicae (SULZER)
Ephedrus (E.) plagiator (NEES)
  Aphis gossypii GLOVER
  Aphis rumicis LINNAEUS [TAKAHASHI, 1925]
  Aphis spiraecola PATCH
  Brachysiphoniella graminis TAKAHASHI [TAKAHASHI, 1925]
  Hyperomyzus lactucae (LINNAEUS)
  Macrosiphum formosanum TAKAHASHI
  Macrosiphum rosaeibarae MATSUMURA
  Megoura citricola V.D. GOOT
  Myzus persicae (SULZER)
  Toxoptera odinae V.D. GOOT
Pauesia unilachni (GAHAN)
  Eulachnus sp.
  Unilachnus sp. [GAHAN, 1926]
Pauesia laticeps (GAHAN)
  Dilachnus sp. [Gahan, 1926]

Trioxys (Binodoxys) carinatus STARÝ AND SCHLINGER
  Macrosiphum rubiformosanum TAKAHASHI
Trioxys (Binodoxys) communis GAHAN
  Aphis gossypii GLOVER [GAHAN, 1926]
  Rhopalosiphum sp.
Trioxys (Binodoxys) indicus SUBBA RAO AND SHARMA
  Aphis gossypii GLOVER
Trioxys (Binodoxys) struma GAHAN
  Megoura citricola (V.D. GOOT) [GAHAN, 1926]
Trioxys (Binodoxys) sp.
  Aphis spiraecola PATCH
Trioxys (Fissicaudus) confucius MACKAUER
  Greenidea ficicola TAKAHASHI
Trioxys (Trioxys) luteolus STARÝ AND SCHLINGER
  Agrioaphis viridis TAKAHASHI
  Shivaphis spp.
Trioxys (Trioxys) sp.
  Phyllaphoides bambusicola TAKAHASHI
Aphidius absinthii MARSHALL
  Macrosiphoniella formosartemisiae TAKAHASHI [GAHAN, 1926]
  Macrosiphoniella tanacetaria KOCH
  "*Macrosiphum* sp."
Aphidius sp. "*gifuensis* ASHMEAD" group
  Capitophorus sp.
  Macrosiphum rosaeibarae MATSUMURA
  Myzus persicae SULZER
  Myzus sp.
Aphidius salicis HALIDAY
  Cavariella araliae TAKAHASHI
Aphidius salignae WATANABE
  Tuberolachnus salignus GMELIN
Aphidius sp. no. 1
  Macrosiphum rosaeibarae MATSUMURA
Praon orientale STARÝ AND SCHLINGER
  Hyperomyzus lactucae LINNAEUS
Lipolexis scutellaris MACKAUER
  Aphis malvoides V.D. GOOT
Lipolexis gracilis FÖRSTER

Aphis spiraecola PATCH
Toxoptera aurantii (B. DE F.)

HONG KONG

Lysiphlebia rugosa STARÝ AND SCHLINGER
  Brachycaudus helichrysi KALTENBACH
Ephedrus (Ephedrus) persicae FROGGATT
  Myzus persicae SULZER
Ephedrus (Ephedrus) plagiator (NEES)
  Cavariella araliae TAKAHASHI
Pauesia unilachni (GAHAN)
  Cinara formosana TAKAHASHI
  Cinara sp.
Trioxys (Binodoxys) sinensis MACKAUER
  Toxoptera sp.
Trioxys (Fissicaudus) confucius MACKAUER
  Greenidea ficicola TAKAHASHI
Praon orientale STARÝ AND SCHLINGER
  Macrosiphum rosaeibarae MATSUMURA
  Myzus sp.
Lipolexis gracilis FÖRSTER
  Aphis gossypii GLOVER
Lipolexis scutellaris MACKAUER
  Unknown aphid
        JAPAN (KYUSHU IS.)
Paralipsis eikoae (YASUMATSU)
  (No host) [YASUMATSU, 1951]
Lysiphlebia japonica (ASHMEAD)
  Aphis spiraecola PATCH
  Toxoptera odinae V.D. GOOT
Diaeretus leucopterus (HALIDAY)
  Eulachnus piniformosanus TAKAHASHI
Ephedrus (E.) persicae FROGGATT
  Myzus sp.
Ephedrus (E.) plagiator (NEES)
  Aphis spiraecola PATCH
  Capitophorus sp.
  Macrosiphum rosaeibarae MATSUMURA
  Toxoptera odinae V.D. GOOT
Pauesia tropicalis STARÝ AND SCHLINGER
  Lachnus tropicalis V.D. GOOT
Bioxys japonicus STARÝ AND SCHLINGER
  Phyllaphidine sp.
Trioxys (Binodoxys) orientalis STARÝ AND
SCHLINGER
  Macrosiphum rosaeibarae MATSUMURA
Aphidius absinthii MARSHALL
  Macrosiphoniella sanborni GILLETTE

Aphidius areolatus ASHMEAD
  Periphyllus testudinatus FERNIE
Aphidius sp., "gifuensis ASHMEAD" group
  Acyrthosiphon sp.
  Macrosiphum sp.
Aphidius salignae WATANABE
  Tuberolachnus salignus GMELIN
Aphidius sp. no. 4
  Rhopalosiphoninus deutzifoliae SHINJI
Praon orientale STARÝ AND SCHLINGER
  Acyrthosiphon sp.
  Amphorophora magnoliae ESSIG AND
  KUWANA
  Aphis spiraecola PATCH
Praon sp. no. 1
  Aphis sp.
Lipolexis gracilis FÖRSTER
  Aphis spiraecola PATCH

        JAPAN (HONSHU IS.)

Diaeretiella rapae (M'INTOSH)
  Brevicoryne brassicae (L.)
Protaphidius nawaii (ASHMEAD)
  Stomaphis yanonis TAKAHASHI [WATA-
  NABE, 1957]
Lysiphlebia japonica (ASHMEAD)
  Aphis gossypii GLOVER [WATANABE,
  1957]
  Aphis spiraecola PATCH
  Brachycaudus helichrysi KALTENBACH
  Parachaitophorus spiraeae TAKAHASHI
  Toxoptera sp.
Diaeretus leucopterus (HALIDAY)
  Eulachnus piniformosanus TAKAHASHI
Ephedrus (Ephedrus) persicae FROGGATT
  Myzus momonis (MATSUMURA)
  Myzus persicae SULZER
Ephedrus (Ephedrus) plagiator (NEES)
  Aphis gossypii GLOVER
  Aphis pomi DE GEER
  Aphis spiraecola PATCH
  Macrosiphum rosaeibarae MATSUMURA
  Parachaitophorus spiraeae TAKAHASHI
  Rhopalosiphoninus deutzifoliae SHINJI
Pauesia japonica (ASHMEAD)
  Lachnus tropicalis V.D. GOOT (?) [WATA-
  NABE, 1939]

Trioxys (Binodoxys) brunnescens STARÝ AND SCHLINGER
   Acyrthosiphon sp.
Aphidius areolatus ASHMEAD
   Periphyllus testudinatus (FERNIE)
Praon orientale STARÝ AND SCHLINGER
   Acyrthosiphon sp.
   Amphorophora sp.
   Macrosiphum rosaeibarae MATSUMURA
   Rhopalosiphoninus deutzifoliae SHINJI
   Rhopalosiphoninus tiliae MATSUMURA
Praon quadratum STARÝ AND SCHLINGER
   Coloradoa rufomaculata WILSON

JAPAN (HOKKAIDO IS.)

Monoctonus (Monoctonus) similis STARÝ AND SCHLINGER
   Myzus sp.
Ephedrus (Ephedrus) persicae FROGGATT
   Myzus mononis (MATSUMURA) [WATA-NABE, 1941]
   Myzus mumecola MATSUMURA [WATANABE, 1941]
Ephedrus (Ephedrus) plagiator (NEES)
   Amphicercus japonicus [WATANABE, 1941]
   Amphorophora magnoliae ESSIG AND KUWANA [WATANABE, 1941]
   Anuraphis mumei [WATANABE, 1941]
   Aphis laburni [WATANABE, 1941]
   Hyalopterus pruni [WATANABE, 1941]
   Macrosiphum rosae LINNAEUS [WATA-NABE, 1941]
   Myzus persicae SULZER [WATANABE, 1941]
   Prociphilus konoi HORI [WATANABE, 1941]
   Sitobium granarium (KIRBY) [WATANABE, 1941]
Pauesia infulata (HALIDAY)
   Cinara sp.
Pauesia pini (HALIDAY)
   Cinara laricicolus (MATSUMURA) [WATA-NABE, 1940]
   Cinara laricis (WALKER) [WATANABE, 1940]
   Cinara pineti (KOCH) [WATANABE, 1940]
Pauesia laricis (HALIDAY)

Cinara laricicolus (MATSUMURA) [WATA-NABE, 1940]
Pauesia inouyei (WATANABE)
   Cinara todocolus (INOUYE) [WATANABE, 1941]
Pauesia jezoensis (WATANABE)
   Cinara nopporoensis INOUYE [WATANABE, 1941]
   Lachniella costata (ZETTERSTEDT) [WATA-NABE, 1941]
Pauesia konoi (WATANABE)
   Cinara longipennis (MATSUMURA) [WA-TANABE, 1941]
Aphidius areolatus ASHMEAD
   Periphyllus koelreuteriae TAKAHASHI
Aphidius sp., "*gifuensis* ASHMEAD" group
   Acyrthosiphon sp.
Aphidius salicis HALIDAY
   Cavariella sp.
Aphidius salignae WATANABE
   Tuberolachnus salignus GMELIN [WA-TANABE, 1939]
Aphidius sp. no. 2
   Acyrthosiphon sp.
Aphidius sp. no. 5
   Acyrthosiphon sp.
Aphidius sp. no. 6
   Amphorophora sp.
Aphidius sp. no. 7
   Acyrthosiphon sp.
Praon glabrum STARÝ AND SCHLINGER
   Euceraphis betulae LINNAEUS
Praon orientale STARÝ AND SCHLINGER
   Amphorophora magnoliae ESSIG AND KUWANA
   Acyrthosiphon sp.

SOUTH KOREA

Lysiphlebia japonica (ASHMEAD)
   Aphis spiraecola PATCH
   Aphis sp.
   Tetraneura sp.
Lysiphlebia rugosa STARÝ AND SCHLINGER
   Aphis sp.
Diaeretus leucopterus (HALIDAY)
   Eulachnus piniformosanus TAKAHASHI
Lysiphlebus salicaphis FITCH

Chaitophorus salicicolus MATSUMURA
Lysiphlebus sp. *aff. delhiensis* SHARMA AND
SUBBA RAO
   Aphis sp.
   Ephedrus (E.) compestris STARÝ
   Macrosiphoniella sanborni GILLETTE
   Macrosiphoniella yomogifoliae SHINJI
   Megoura viciae BUCKTON
Ephedrus (E.) persicae FROGGATT
   Aphis gossypii GLOVER
   Myzus persicae SULZER
Ephedrus (E.) plagiator (NEES)
   Amphorophora lonicericola TAKAHASHI
   Aphis spiraecola PATCH
Ephedrus (Lysephedrus) validus (HALIDAY)
   Eriosomatinae sp.
Pauesia unilachni (GAHAN)
   Cinara orientalis TAKAHASHI
Trioxys (Binodoxys) orientalis STARÝ AND
SCHLINGER

   Macrosiphum rosaeibarae MATSUMURA
Aphidius absinthii MARSHALL
   Macrosiphoniella formosartemisiae TAKA-
HASHI
   Macrosiphoniella sanborni GILLETTE
   Macrosiphoniella yomogifaliae TAKA-
HASHI
Aphidius sp., "*gifuensis* ASHMEAD" group
   Myzus persicae SULZER
Aphidius salicis HALIDAY
   Cavariella salicicola (MATSUMURA)
Aphidius sp. no. 3
   Macrosiphum rosae LINNAEUS
Praon orientale STARÝ AND SCHLINGER
   Aphis spiraecola PATCH
   Cavariella sp.
   Macrosiphum rosaeibarae MATSUMURA
Areopraon sp.
   Periphyllus sp.

REFERENCES

ASHMEAD, W. H. 1906. Descriptions of new Hymenoptera from Japan. *Proc. U. S. Nat. Mus.* 30: 169-201, Pl. XI-XV.

BAKER, C. F. 1909. Plant louse parasites I. *Pomona J. Ent.* 1: 22-25.

BARNES, H. F. 1931. Notes on the parasites of the cabbage aphid (*Brevicoryne brassicae* Lin.). *Ent. mon. Mag.* 67: 55-7.

BEARDSLEY, J. W. 1961. A Review of the Hawaiian Braconidae. *Proc. Hawaiian Ent. Soc.* 17(3):333-366.

BEIRNE, B. P. 1942. Observations on the developmental stages of some Aphidiinae. *Ent. mon. Mag.* 78: 283-6, 4 figs.

BLANCHARD, E. E. 1940. Descripción de un nuevo Afidiino argentino, util para la agricultura. *Rev. Chil-Hist. nat. Santiago.* 44: 45-8, 1 p.

BRITTON, W. E. 1917. A destructive aphid on turnip, *Aphis pseudobrassicae* Davis. 16th Rept. State Entomologist of Connecticut for 1916, Conn. Agric. Expt. Sta., New Haven. pp. 98-104.

DALLA TORRE, K. W. VON. 1898. Catalogus Hymenopterorum huiusque descriptorum systematicus et synonymicus. Vol. IV. Braconidae. Lipsiae.

ENTOMOPHAGA. 1960. Liste d'Identif. No. 3. *Entomophaga* 5: 340

FAHRINGER, J. 1934. Hym. Braconidae, in Schwedisch-Chinesische wissenschaftliche Expedition nach den nordwestlichen Provinzen Chinas. *Ark. f. Zool.* 27A: 1-15.

— 1937. Die Parasiten der Baumläuse (Lachnini) aus der Gruppe der Aphidiinae Först. Festschr. 60. Geb. E. Strand, Riga. 3: 240-5.

FERRIÈRE, C. and VOUKASSOVITCH, P. 1928. Sur les parasites des aphids et leurs hyperparasites. *Bull. Soc. ent. Fr., Paris.* 1928: 26-9.

FLANDERS, S. E. and FISHER, T. W. 1959. The economic effect of aphidophagous insects on citrus in South China. *J. econ. Ent.* 52: 536-7.

FÖRSTER, A. 1862. Synopsis der Familien und Gattungen der Braconen. *Verh. naturh. Ver. Preuss. Rheinl.* 19: 225-50.

FROGGATT, W. W. 1904. Experimental work with the peach aphis (*Aphis persicae-niger*, Sm.). *Agric. Gaz. N. S. Wales*, 15: 603-612.

FULLAWAY, D. T. 1915. Report of the Entomologist. Rept. Hawaii Agric. Expt. Sta. 1914, Washington, 43-50.

GAHAN, A. B. 1910. Some synonymy and other notes on Aphidiidae. *Proc. Ent. Soc. Washington.* 12: 179-89.

— 1917. Descriptions of some new parasitic Hymenoptera. *Proc. U. S. Nat. Mus.* 53, No. 2197: 195-217.

— 1926. Some braconid and chalcid flies from Formosa, parasitic on aphids. *Proc. U. S. Nat. Mus.* 70(8): 1-7.

— 1932. Miscellaneous descriptions and notes on parasitic Hymenoptera. *Ann. Ent. Soc. Amer.* 25: 736-57.

GAUTIER, A. and BONNAMOUR, S. 1929. Remarques sur *Ephedrus plagiator* Nees, complément de description de cette espèce. *Bull. Soc. ent. Fr.* 1929: 92-5.

— 1936. Sur deux *Aphidius* nouveaux du pin. *Bull. Soc. linn. Lyon* N.S. 5(5): 74-5.

GAUTIER, A., BONNAMOUR, S. and GAUMOND L. 1929. Sur un *Ephedrus* (Hym. Brac.) parasite de *Macrosiphoniella sanborni* Gillette. *Bull. Soc. ent. Fr.* 1929: 200-1.

GEORGE, K. S. 1957. Preliminary investigations on the biology and ecology of the parasites and predators of *Brevicoryne brassicae* L. *Bull. ent. Res.* 48: 619-29.

GILMORE, J. E. 1960. Biology of the black cherry aphid in the Willamette Valley, Oregon. *J. econ. Ent.* 53(4): 659-61.

GOIDANICH, A. 1934. Materiali per lo studio degli Imenotteri Braconidi. II. *Boll. Lab. Ent. Bologna.* 6: 209-30.

GOURLAY, E. S. 1930. Preliminary host-list of the entomophagous insects in New Zealand. *Bull. N.Z. Dept. Sci. Industr. Res.* 22: 13 pp.

GRANGER, C. 1949. Braconides de Madagascar. *Mem. Inst. Sci. Madagascar* A(2): 1-428.

HAGEN, K. S. and E. I. SCHLINGER. 1960. Imported Indian parasite of pea aphid established in California. *Calif. Agric.* 14(9): 5-6.

HALIDAY, A. D. 1833. An essay on the classification of the parasitic Hymenoptera of Britain, which correspond with the Ichneumones minuti of Linnaeus. *Ent. Mag.* 1: 259-76 and 480-91.

— 1834. Essay on the classification of parasitic Hymenoptera of Britain, which correspond with the Ichneumones minuti of Linnaeus. *Ent. Mag.* 2: 93-106.

HALL, J. C., SCHLINGER, E. I. and R. VAN DEN BOSCH. 1962. Evidence for the separation of the "Sibling Species" *Trioxys utilis* and *Trioxys pallidus*. *Ann. Ent. Soc. Amer.* 55: 566-68.

HAVILAND, M. D. 1921. On the bionomics and development of *Lygocerus testaceimanus* Kieffer and *Lygocerus cameroni* Kieffer (Proctotrypoidea, Ceraphronidae), parasites of *Aphidius*. *Quart. J. microsc. Sci.* 65: 101-27.

— 1921. On the bionomics and post-embryonic development of certain cynipid hyperparasites of aphids. *Quart. J. microsc. Sci.* 65: 451-78.

— 1922. On the post-embryonic development of certain chalcids, hyperparasites of aphides, with remarks on the bionomics of hymenopterous parasites in general. *Quart. J. microsc. Sci.* 66: 321-38.

HERRICK, G. W. and HUNGATE, J. W. 1911. The cabbage aphis (*Aphis brassicae* L.). *Cornell Univ. Agric. Expt. Sta. Coll. Agric. Bull.* 300: 717-46.

HINCKS, W. D. 1949. A genus and species of aphidiid new to Sweden from Linné's garden at Hammarby. *Ent. Tidskr.* 70: 171-4, 1 fig.

— 1958. *Myrmecobosca mandibularis* Maneval (Hym. Braconidae), a myrmecophilous aphid parasite in Britain. *Ent. mon. Mag.* 94: 20-1, 1 fig.

HODEK, I., STARÝ, P. and P. STYS. 1962. The natural enemy complex of *Aphis fabae* and its effectiveness in control. XI. Int. Cong. Ent. Wien. 1960, Bd. 2, pp. 747-749.

IVANOVA–KASAS, O. M. 1956. Vergleichende Studien an Embryonal-Entwicklung von *Aphidius* and *Ephedrus*. *Rev. d'Ent. de l'URSS* 35: 245-61.

— 1961. Otscherki po sravnitelnoi embryologii perepontshatokrylych. Leningrad, 265 pp.

IWATA, K. 1959. The comparative anatomy of the ovary in Hymenoptera. Part III: Braconidae (including Aphidiidae) with descriptions of ovarian eggs. *Kontŷu* 27(4): 231-8, Pl. 15-16.

KIRCHNER, L. 1867. Catalogus Hymenopterorum Europae. Vindobonae, 285 pp.

KLOET, G. S. and HINCKS, W. D. 1945. A check list of British insects. Stockport, 483 pp.

KROMBEIN, K. V., et al. 1958. Hymenoptera of America north of Mexico. Synoptic Catalog U.S. Dept. Agric. Monogr. 2, First Suppl., 305 pp.

KURDJUMOV, N. 1913. One new aphid feeding braconid. *Rev. russ. ent. St. Petersburg* 13 : 25-6.

LINDROTH, C. H. 1931. Die Insektenfauna Islands und ihre Probleme. *Zool. Bidrag, Uppsala* 13 : 105-599.

LUZHETSKI, A. N. 1959. Fauna parazitov sem. Aphidiidae Uzbekistana. Tezisy Dokl. 4-ogo sjezda Vses. Entom. Obstschestva, AN SSSR, Moskva, Leningrad, pp. 82-3.

— 1960. Faunistitheskije osnovy ispolzhovanija entomophagov dlja biologitsheskoj borby s selsko-khozjastvennymi vrediteljami v Uzbekistane. Nautsh. Issl. po zasth. rastenij, Tashkent, 247-50.

— 1960. Parazity tlej Uzbekistana, pp. 89-163, in: YAKHONTOV V. V., LUZHETSKI, A. N., ALIMDZHANOV, R. A. 1960: Poleznye i vrednye nasekomye Uzbekistana. Izd. AN Uzb. SSR, Tashkent, 1-201 pp.

MACGILLIVRAY, M. E. and SPICER, P. B. 1953. Aphid parasites collected in New Brunswick in 1950. *Canad. Ent.* 85(11): 423-31.

MACKAUER, M. 1959. Die mittel-, west- und nordeuropäischen Arten der Gattung Trioxys Haliday. *Beitr. Ent.* 9: 144-79, 39 figs.

— 1959. Die systematische Stellung von *Aphidius pseudoplatani* Marsh. *Senck. biol. Frankfurt M.*, 40: 179-82, 8 figs.

— 1959. Die europäischen Arten der Gattungen *Praon* und *Areopraon*. *Beitr. Ent.* 9: 810-65, 38 figs.

— 1960. Zur Systematik der Gattung *Trioxys* Haliday. *Beitr. Ent.* 10: 137-60.

— 1960. Zur Kenntnis der nearktischen Arten der Gattung *Lysiphlebus* Förster. *Boll. Lab. Ent. Agr. Portici.* 18: 230-56, 2 figs.

— 1960. *Trioxys (Trioxys) vandenboschi* n. sp., ein neuer Blattlaus – Parasit aus dem Iran. *Senck. biol., Frankfurt M.*, 41(5-6): 359-62, 4 figs.

— 1960. Die europäischen Arten der Gattung *Lysiphlebus* Förster. *Beitr. Ent.* 10: 582-623, 29 figs.

— 1961a. Die Typen der Unterfamilie Aphidiinae des Britischen Museums London. *Beitr. Ent.* 11: 96-154.

— 1961b. Neue europäische Blattlaus-Schlupfwespen. *Boll. Lab. Ent. Agr. Portici*, 19: 270-90.

— 1961c. Zur Frage der Wirtsbindung der Blattlaus-Schlupfwespen. *Z. Parasitenkunde* 20: 576-91.

— 1962a. Blattlaus-Schlupfwespen der Sammlung F. P. Müller, Rostock. *Beitr. Ent.* 12(5/6): 631-661.

— 1962b. Drei neue Blattlaus-Parasiten aus Hong Kong. *Entomophaga* 7(1):37-45, figs. 1-8.

— 1962c. Spezifische Parasiten der *Acyrthosiphon-Macrosiphum*-Gruppe und Grundfragen der Wirtsbindung der Blattlaus-Schlupfwespen. *Z. ang. Ent.* 50(1): 125-131.

— 1963. Bemerkungen zur Systematik, Verbreitung und Wirtsbindung des *Ephedrus persicae*-Komplexes. *Z. ang. Ent.* 52(4): 343-354.

MANEVAL, H. 1940. Observations sur un Aphidiidae myrmécophile. Description du genre et de l'espèce. *Bull. Soc. linn. Lyon* 9: 9-14, 7 figs.

MARSHALL, T. A. 1891, 1897. Braconides, in ANDRÉ, Spécies des Hyménoptères d'Europe et d'Algérie. Vol. V, V bis.

— 1899. A monograph of the British Braconidae. Pt. VIII (Flexiliventres). *Trans. ent. Soc. London* 1899: 11-76.

MASON, A. C. 1922. Life-history studies of some Florida aphids. *Florida Ent.*, 5(4): 53-9, 62-5.

MELANDER, A. L. and YOTHERS, M. A. 1915. Twenty-fifth and twenty-sixth annual reports of the years ended 30th June 1915 and 30th June 1916, Division of Zoology and Entomology. *Washington State Coll. Agric. Expt. Sta., Bull.* 127: 30-38, 136: 35-42.

MESSENGER, P. S. and FORCE, D. C. 1963. An experimental host parasite system: *Therioaphis maculata* (Buckton)-*Praon palitans* Muesebeck. *Ecology* 44: 532-40.

MILLAN, E. 1956. Metamorfosis y ecología de *Aphidius platensis* Brethes. *Rev. Invest. Agric.* 10: 243-280.

MUESEBECK, C. F. W. and WALKLEY, L. M. 1951. in MUESEBECK, C. F. W., KROMBEIN, K. V. and TOWNES, H. K. 1951. Hymenoptera of America North of Mexico. Synoptic Catalog. U.S. Dept. Agric. Monogr. 2, Washington.

MUKERJI, S. 1948. *Aphidius antennatus* sp. nov., parasitic on *Pterochlorus persicae* Chlodk., affecting *Prunus persica* (Peach) in Baluchistan. *Indian J. agric. Sci.* 18: 33-4.

MUKERJI, S. and CHATTERJEE, S. N. 1950. *Diaeretus aphidum* sp. n., a parasite of *Pterochlorus persicae* Cholodkovsky on *Prunus persica* in Baluchistan. *Proc. R. ent. Soc. London* (B)19: 4-6, 8 figs.

NARAYANAN, E. S., SUBBA RAO, B. R. and SHARMA, A. K. 1960. A catalogue of the known species of the world belonging to the subfamily Aphidiinae. *Beitr. Ent.* 10(5-6): 545-581.

NARAYANAN, E. S., SUBBA RAO, B. R., SHARMA, A. K. and STARÝ, P. 1962. Revision of "A catalogue of the known species of the world belonging to the subfamily Aphidiinae". *Beitr. Ent.* 12(5/6): 662-720.

NARZYKULOV, M. N. and ATAEVA, M. A. 1961. Materialy k faune najezdnikov (Hymenoptera, Aphidiidae), parazitov tlej v Srednej Azii. *Trudy In-ta zool. i parazitol. AN Tadzh. SSR* 20: 189-91.

NEES AB ESENBECK, C. G. 1819. Appendix ad Gravenhorst conspectum generum et familiarum Ichneumonidum, genera et familias Ichneumonidum adscitorum exhibens. *Verh. Leop.-Carol. Acad. Naturforsch., Erlangen* 9: 299-310.

— 1834. Hymenopterorum Ichneumonibus affinium monographiae, genera europaea et species illustratae. Stuttgartiae et Tubingiae.

NIEZABITOWSKI, E. L. 1909. Materyaly do fauny Brakonidów Polski. *Spraw. Kom. Fiziogr. Kraków* 44: 47-106.

PETERSEN, B. 1965. The Zoology of Iceland. III Vol., Pt. 49-50. Copenhagen and Reykjavík.

PETHERBRIDGE, F. R. and MELLOR, J. E. M. 1936. Observations on the life history and control of the cabbage aphis, *Brevicoryne brassicae* L. *Ann. appl. Biol.* 23: 329-41, Cambridge.

PONTIN, A. J. 1960. Some records of predators and parasites adapted to attack aphids attended by ants. *Ent. mon. Mag.* 95: 154-5.

QUILIS, M. P. 1931. Especies nuevas de Aphidiidae españoles. *Eos* 7: 25-84, 98 figs.

— 1934. Algunos Aphidiidae de Checoslovaquia. *Eos* 10: 5-19, 17 figs.

REGNIER, R. 1923. De quelques grands ennemis du pommier et de leurs parasites. *Rev. Bot. appl. Agric. colon.*, III (19): 169-85.

RIPPER, W. E. 1944. Biological control as a supplement to chemical control of insect pests. *Nature Lond.* 153, No. 3885: 448-52, 2 figs.

SAVARY, A. 1953. Le puceron cendré du poirier (*Sappaphis pyri* Fonsc.) en Suisse romande. *Landw. Jb. Schweiz* 67: 247-314, 2 pls., 13 figs.

SCHIMITSCHEK, E. 1935. Forstschädlingsauftreten in Österreich 1927 bis 1933. *Cbl. ges. Forstwesen* 61: 134-221.

— 1936. Ergebnisse von Parasitenzuchten. *Z. ang. Ent.* 22: 558-564.

— 1936. Forstschädlingsauftreten in Österreich 1934 und 1935. *Cbl. ges. Forstwes.* 62: 65-120.

SCHLINGER, E. I. 1960. Diapause and secondary parasites nullify the effectiveness of the rose aphid parasites in Riverside, California. *J. econ. Ent.* 53: 151-4.

SCHLINGER, E. I., HAGEN, K. S. and R. VAN DEN BOSCH. 1960. Imported French parasite of walnut aphid established in California. *Calif. Agric.* 14(11): 3-4.

SCHLINGER, E. I. and HALL, J. C. 1959. A synopsis of the biologies of three imported parasites of the spotted alfalfa aphid. *J. econ. Ent.* 52: 154-7.

— 1960a. Biological notes on pacific coast aphid parasites, and lists of California parasites (Aphidiinae) and their aphid hosts. *Ann. Ent. Soc. Amer.* 53: 404-15.

— 1960b. The biology, behavior and morphology of *Praon palitans* Muesebeck, an internal parasite of the spotted alfalfa aphid, *Therioaphis maculata* (Buckton). *Ann. Ent. Soc. Amer.* 53: 144-160, figs.

— 1961. The biology, behavior and morphology of *Trioxys (Trioxys) utilis*, an internal parasite of the spotted alfalfa aphid, *Therioaphis maculata*. *Ann. Ent. Soc. Amer.* 54: 34-45, figs.

SCHLINGER, E. I. and M. J. P. MACKAUER. 1963. Identity, distribution, and hosts of *Aphidius matricariae* Haliday, an important parasite of the green peach aphid, *Myzus persicae*. *Ann. Ent. Soc. Amer.* 56: 648-53.

SEDLAG, U. 1958. Beobachtungen über das Auftreten der Kohl-blattlaus (*Brevicoryne brassicae* L.) im Sommer 1957. *Nachrbl. Deutsch. Pflanzenschutzdienst* 12: 73-77.

— 1959. Untersuchungen über Bionomie, Anatomie und Massenwechsel von *Diaeretus rapae* Curt. (Hym. Aphidiidae). (Autoreferat einer Habilitationsschrift). *Wiss. Z. E.M. Arndt-Univ. Greifswald* Jahrg. 7, 1957/58, Math.-naturw. Reihe, 3/4, 1-2 pp.

— Untersuchungen über Bionomie und Massenwechsel von *Diaeretus rapae* (Curt.). Trans. I. Int. Conf. Insect Pathology and Biol. Control, Praha (1958): 367-72.

SEITNER, M. 1936. *Lachnus cembrae* n. sp. Die Zirbenblattlaus. *Zbl. Forstwes.* 62(2): 33-49, 8 figs. Vienna.

SHANDS, W. A., SIMPSON, G. W. ROBERTS, F. S. and MUESEBECK, C. F. W. 1955. Parasites of potato – infesting aphids and of some other aphids in Maine. *Proc. ent. Soc. Wash.* 57: 131-36.

SILVEIRA, GUIDO A. and RUFFINELLI, A. 1958. Primer Catalogo de los Parasitos y predatores encontrados en el Uruguay. Proc. 10th Int. Congr. Ent. Montreal, 1956, 4, 913-24.

SKRIPTSHINSKIJ, G. 1930. Zur Biologie von *Aphidius granarius* Marsh and *Ephedrus plagiator* Nees (Braconidae), Parasiten von *Aphis padi* L. *Rep. Appl. Ent.*, 4: 351-64, 14 figs.

SMITH, C. F. 1944. The Aphidiinae of North America. *Ohio State Univ. Contr. Zoo. Ent.*, 6, 154 pp., 17 pls.

SPENCER, H. 1926. Biology of the parasites and hyperparasites of aphids. *Ann. Ent. Soc. Amer.* 12(2): 119-57.

STARKE, H. 1956. Ichneumonidenfauna der sächsischen Oberlausitz. *Natura Lusatica*, 3, 17-92.

STARÝ, P. 1956. Notes on the Braconidae of Czechoslovakia. II. *Acta Mus. Siles. Opava* 5: 47-8.

— 1958. A taxonomic revision of some aphidiine genera with remarks on the sub-family Aphidiinae. *Acta Faun. Ent. Mus. Nat. Pragae* 3: 53-96, 12 Tbls.

— 1959a. Redescription of the aphidiine genus *Lipolexis* Förster. *Acta Soc. Ent. Čechosl.* 56: 93-6, 6 figs.

— 1959b. A revision of the European species of the genus *Monoctonus* Haliday. *Acta Soc. Ent. Čechosl.* 56: 237-50.

— 1959c. Synonymical and other notes on *Protaphidius wissmannii* (Ratzeburg). *Ins. Mats.*, 22: 88-91.

— 1960a. The systematic position of "*Diaeretus oregmae* Gahan". *Ins. Mats.* 23(2): 109-11.

— 1960b. Une nouvelle espèce du genus Trioxys Hal. de la région de la Pannonie (Tchécoslovaquie). *Bull. Soc. Ent. Mulhouse* 1960. 93-6, 3 figs.

— 1960c. A taxonomic revision of the European species of the genus *Paraphidius* Starý 1958. *Acta Faun. Ent. Mus. Nat. Pragae* 6: 5-55, 64 figs.

— 1960d. The generic classification of the family Aphidiidae. *Acta Soc. Ent. Čechosl.* 57: 238-52, 13 figs.

— 1961a. A revision of the genus *Diaeretiella* Starý. *Acta Ent. Mus. Nat. Pragae* 34: 383-97, 4 figs.

— 1961b. Taxonomic notes on the genus *Lysiphlebus* Förster. *Bull. ent. Pologne* 31(9): 96-103, 10 figs.

— 1961c. Notes on the parasites of the root aphids. *Acta Soc. Ent. Čechosl.* 58(3): 228-38, 7 photogr.

— 1961d. Some synonymical notes on the Aphidiidae. *Acta Ent. Mus. Nat. Pragae* 34: 21-5, 2 figs.

— 1961e. Notes of European species of the genus *Aphidius* Nees. *Ent. Tidskr.* 82: 213-21

— 1961f. Faunistic survey of Czechoslovak species of the genera *Lysiphlebus* Förster and *Trioxys* Haliday. *Acta Faun. Ent. Mus. Nat. Pragae* 7: 131-49.

— 1961g. Notes on the Aphidiidae of Crimea. *Acta Faun. Ent. Mus. Nat. Pragae* 7: 129-30.

— 1962a. Aphidofauna of honey plants as a source of subsidiary hosts of aphidiid wasps. *Acta Soc. Ent. Čechosl.* 59(1): 42-58.

— 1962b. Faunistic notes on the Aphidiidae of Bulgaria. *Acta Faun. Ent. Mus. Nat. Pragae* 8(71): 83-86.

— 1962c. Bionomics and ecology of *Ephedrus pulchellus* Stelfox an important parasite of leaf-curling aphids in Czechoslovakia, with notes on the diapause. *Entomophaga* 7(2): 91-100.

— 1962d. Hymenopterous parasites of the pea aphid *Acyrthosiphon onobrychis* (Boyer) in Czechoslovakia. I. Bionomics and ecology of *Aphidius ervi* Haliday. *Zoologicke Listy, Folia Zoologica* 11(15): 265-278.

— 1962e. Notes on aphid parasites of the southern Crimea. *Revue d'Ent. URSS.* 41(4): 875-877, fig. 1-5.

— 1962f. Notes on the European species of the Genus *Ephedrus* Haliday. *Opusc. entomol.* 27: 87-98.

— 1963a. A study of the relationship of the Myzinae and their aphidiid parasites in (Central) Europe. *Boll. Lab. Agr. Portici* 21: 199-216.

— 1963b. A study on the relationship of the Lachnidae, Chaitophoridae, Thelaxidae,

Eriosomatidae, Adelgidae, Phylloxeridae and their aphidiid parasites in (Central) Europe. *Beitr. Ent.* 13(7/8): 894-901.

— 1963c. A study on the relationship of the Dactynotinae and their aphidiid parasites in Europe. *Acta Ent. Mus. Nat. Pragae* 35: 593-610.

— 1963d. Food specificity in the Aphidiidae. *Entomophaga* 9(1): 91-99.

STARÝ, P. and A. RUPAIS 1963. The parasites of dendrophilous aphids in East Baltic. *Latvijas Entomologs* 7: 63-67.

STELFOX, A. W. 1941. A list of the Irish species of *Toxares* and *Ephedrus* (Hym., Aphidiidae) with descriptions of these, which include three species new to science. *Proc. R. Irish Acad., Dublin* 46(B): 125-42, 25 figs.

— 1944. Notes on *Ephedrus*. *Ent. mon. Mag.*, 80: 235-6.

— 1957. Abundance of *Ephedrus validus* Hal. (Hym., Aphidiidae) in September, 1955. *Ent. mon. Mag.* 93. p. 91.

STRICKLAND, E. H. 1916. Control of cabbage aphis by parasites in western Canada. *Proc. B. C. Entom. Soc. Victoria*, 1916. Entom. Series No. 9: 84-8, 3 figs.

SUBBA RAO, B. R. and SHARMA, A. K. 1958. *Trioxys (Trioxys) indicus*, new species, a parasite of *Aphis gossypii* Glov. *Indian J. Ent.* 20: 199-202.

— 1960. Three new species of Braconidae from India. *Proc. Indian Acad. Sci.* 51: 82-8.

— 1962. Studies on the biology of *Trioxys indicus* Subba Rao and Sharma 1958, a parasite of *Aphis gossypii* Glover. *Proc. Nat. Inst. Sci. India* 28, B(2): 164-182. 2 text figs.

SZÉPLIGETI, G. V. 1904. Braconidae, in Wytsman Genera insectorum Fasc. 22. Bruxelles.

TAKAHASHI, R. 1925. Aphididae of Formosa. Part. 4. Dept. Agric. Govt. Res. Inst., Formosa, Rept. 16, 74 pp.

TELENGA, N. A. 1950. K voprosu ob ispolzovanii parazitov sem. Aphidiidae v borbe s migrirujusthimi tljami. Nautsh. *Trudy In-ta ent. filtop. AN Ukr. SSR, Kiev*, 2: 199-209.

— 1953. Novye vidy parazitov tlej Uzbekistana. *Trudy In-ta Zool. parazitol. AN Uzb. SSR, Tashkent*, 1: 169-73.

THUNEBERG, E. 1960. Beiträge zur Kenntnis der finischen Blatt und Schildläuse (Hom. Aphidoidea et Coccoidea) sowie deren Parasiten. *Suomen hyönteistieteelll. Aikakauskirja* 26(1): 97-9.

TIMBERLAKE, P. H. 1918. Notes on some of the immigrant parasitic Hymenoptera of the Hawaiian Islands. *Proc. Hawaiian Entom. Soc.* for 1917, Honolulu, 3(5): 399-404.

TODD, D. H. 1957. Incidence and parasitism of insect pests of cruciferous crops in Hawke's Bay, Wairarapa, and Manawatu, 1955-56 *N.Z. J. Sci. Tech.* 38(A): 720-7.

TREMBLAY, E. 1964. Ricerche sugli imenotteri parassiti. I.-Studio morfo–biologico Sul *Lysiphlebus fabarum* (Marshall). *Boll. Lab. Ent. Agr. Portici*, 22: 1-122, figs.

TREHERNE, R. C. 1916. A preliminary list of parasitic insects known to occur in Canada. 46th Ann. Rept. Dept. Entomol. Soc. Ontario, 1915, Toronto, 1916, 178-93.

VAN DEN BOSCH, R. 1957. The spotted alfalfa aphid and its parasites in the Mediterranean Region, Middle East, and East Africa. *J. econ. Ent.* 50: 352-6.

VAN DEN BOSCH, R. and E. I. SCHLINGER. 1965. A density dependent action in nature and factors tending to obscure its expression. Trans. 12th Int. Cong. Entom. 1964:379.

VAN DEN BOSCH, R., SCHLINGER, E. I., DIETRICK, E. J., HAGEN, K. S. and J. K. HOLLOWAY. 1959a. The colonization and establishment of imported parasites of the spotted alfalfa aphid in California. *J. econ. Ent.* 52: 136-41.

VAN DEN BOSCH, R., SCHLINGER, E. I., DIETRICK, E. J., and I. M. HALL. 1959b. The role of imported parasites in the biological control of the spotted alfalfa aphid in southern California. *J. econ. Ent.* 52: 142-54.

VAN DEN BOSCH, R., SCHLINGER, E. I., DIETRICK, E. J., HALL, J. C. and B. PUTLER. 1964. Studies on succession, distribution, and phenology of imported parasites of *Therioaphis trifolii* (Monell) in southern California. *Ecology*, 45(3): 602-21.

VIERECK, H. T. 1911. Descriptions of six new genera and thirty-one new species of Ichneumon flies. *Proc. U. S. Nat. Mus.* 40: 170-96.

— 1912. Descriptions of five new genera and twenty-six new species of Ichneumon flies. *Proc. U. S. Nat. Mus.* 42: 39-153.

— 1914. Type species of the genera of Ichneumon flies. *Bull. U. S. Nat. Mus.* 83: 1-186.

VEVAI, R. J. 1942. On the bionomics of *Aphidius matricariae* Hale, a braconid parasite of *Myzus persicae* Sulz. *Parasitology* 34: 141-151.

VUKASOVIČ, P. 1928. Contribution to the study of entomophagous insect parasites. *Glas Srpske Kral. Akad.* 131: 45-72, Belgrade.

WATANABE, C. 1939. A new species of genus *Aphidius* Nees and re-description of *Aphidius japonicus* Ashmead. *Ins. Mats.*, 13: 81-4, 1 fig.

— 1940. On two species of *Aphidius* bred from *Cinara laricicolus* (Matsumura). *Ins. Mats.* 15: 53-6, 1 fig.

— 1941. On two aphidiid parasites of the grain aphis, *Macrosiphum granarium* (Kirby). *Ins. Mats.*, 15: 167-70.

— 1941. Descriptions of three new species of *Aphidius* parasitic on some aphids of coniferous trees. *Ins. Mats.* 15: 106-11, 2 figs.

— 1941. On two species of genus *Ephedrus* Haliday. *Ins. Mats.* 15: 136-40.

— 1948. Evaniidae, Gasteruptionidae and Aphidiidae from Shansi, China. *Mushi* 19. 31-2.

— 1949. Aphidiidae of Inner Mongolia. *Mushi* 20: 43-5.

— 1957. Notes on Ashmead's Japanese Braconidae. *Ins. Mats.* 21(1/2): 1-5.

WATANABE, CHIHISA and HAJIMU TAKADA. 1964a. Occurrence of two species of the genus *Praon* Haliday in Japan. *Ins. Mats.* 27(1): 8-11.

— 1964b. A note on *Pauesia konoi* (Watanabe). *Ins. Mats.* 27(1): 11.

WESMAEL, C. 1835-8. Monographie des Braconides de Belgique. *Nouv. Mém. Acad. Bruxelles* IX-XI.

WHEELER, E. W. 1923. Some braconids parasitic on aphids and their life-history. *Ann. Ent. Soc. Amer.* 16: 1-29, 1 tbl., 9 figs.

WIACKOWSKI, S. K. 1962. Studies on the biology and ecology of *Aphidius smithi* Sharma and Subba Rao, a parasite of the pea aphid, *Acyrthosiphon pisum* (Harr.). *Bull. Ent. Pologne* 32(21): 253-310, figs. 1-11.

WIACKOWSKI, S., and WIACKOWSKA, I. 1961. Results of cultivating parasites of orchard entomofauna. Part II. *Bull. ent. Pologne* 31(18): 255-62.

YASUMATSU, K. 1951. Sur une remarquable Aphidiide du Japon. *Rev. franç. Ent.* 18: 171-4, 1 fig.

— 1960. The identity of *Paralipsis enervis* (Nees) and *P. eikoae* (Yasumatsu). *Kontyu* 28: 57.

YASUMATSU, K., ISHIHARA and MORITSU 1946. Five hymenopterous parasites of *Tuberolachnus saligna* Gmelin (Aphididae) from Japan. *Mushi* 17: 9-12.

PLATES

Plate I. Morphological features of head and abdomen

1. Anterior view of head. OC–ocellus, FR–frons, O–eye, FC–frontoclypeus, TE–temple, GE–gena, MD–mandible, PMX–maxillary palpus, PLB–labial palpus, LBR–labrum, CL–clypeus.
2. Anterior view of head showing numbered lines used in head measurements. Line 1–interocular, 2–facial, 3–head width, 4–transfacial, 5–intertentorial, 6–tentorio-ocular, 7–width of gena, 8–socket diameter, 9–socket ocular, 10–vertical eye diameter, 11–longitudinal eye diameter.
3. Enlarged anterior view of mouth area. CL–clypeus, O–eye, MD–mandible, LBR–labrum.
4. Posterior view of head. CO–occipital suture, OC–ocellus, VX–vertex, OCC–occiput, FO–occipital foramen, TE–temple.
5. Lateral view of abdomen.

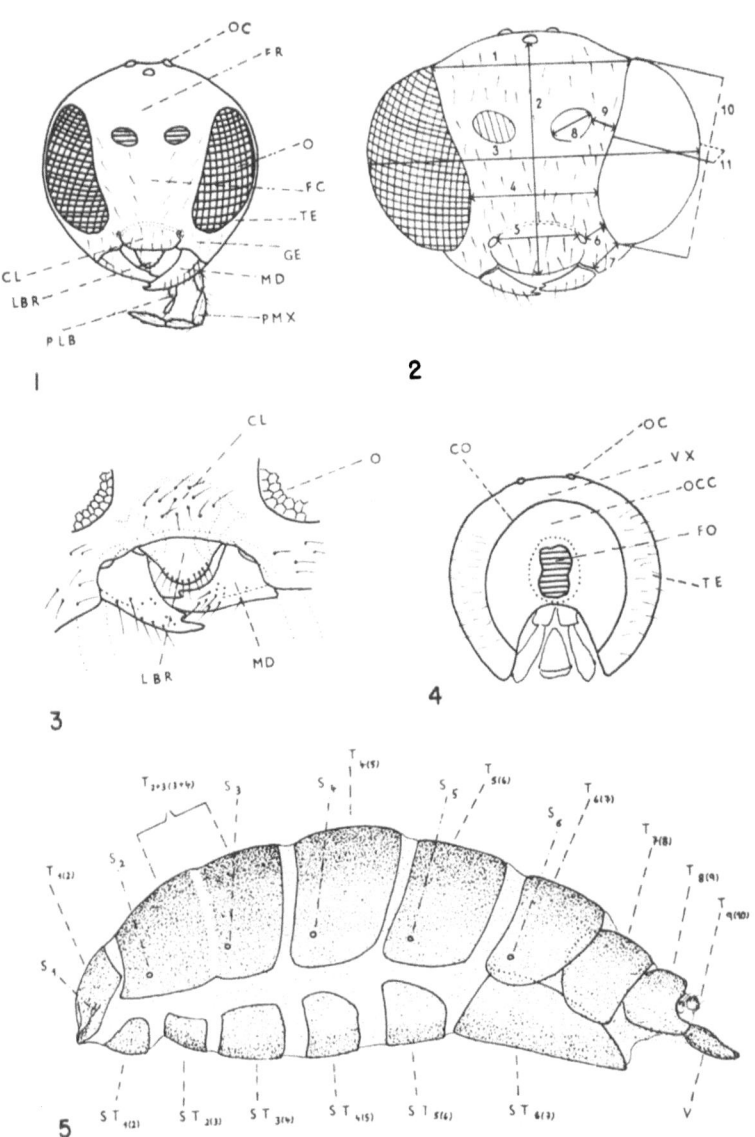

Plate II. Morphological features of wing and thorax

6. Forewing and hindwing. Veins: C+Sc−costa+subcosta, Pt-pterostigma, Mt-metacarpus, R-radius, M-medius, B-basal, Cu-cubitus, Im-intermedian, Ir-inter-radial; cells named according to veins.
7. Thorax and anterior part of abdomen in dorsal view. PN-pronotum, NOT-notaulices, MSCT-mesoscutum, TG-tegula, AANT-forewing, AX-axilla, SCTL-scutellum, MTN-metanotum, APST-hindwing, PSCTL-postscutellum, PROP-propodeum, SP-spiracle, T_1-tergite I, T_2-tergite 2.

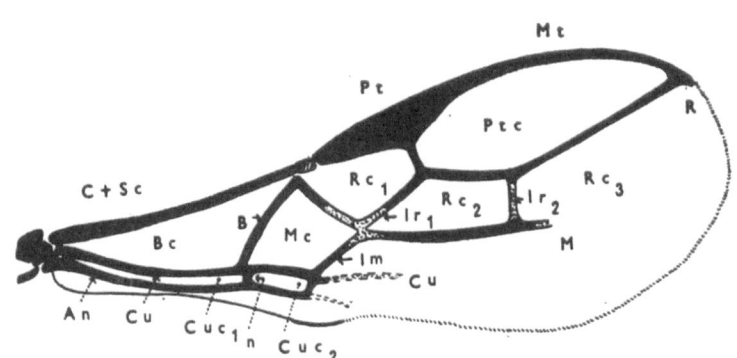

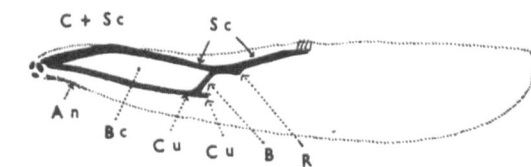

6

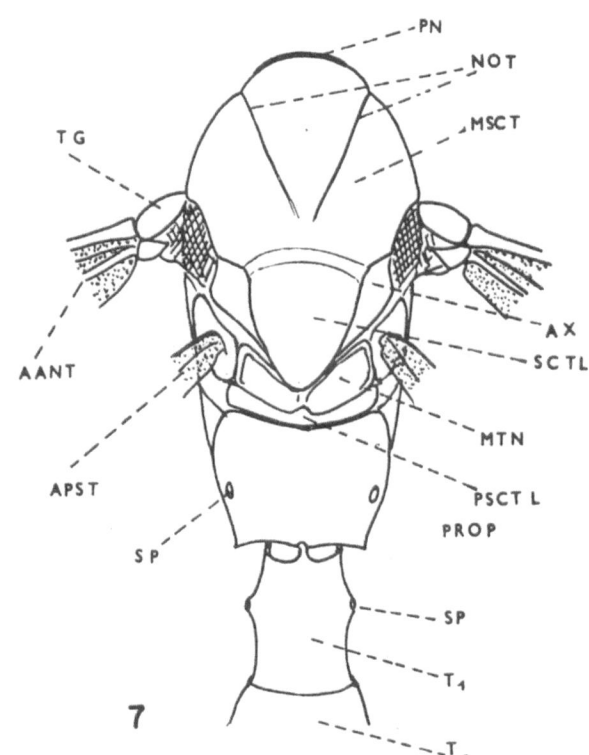

7

Plate III. Morphological features of genitalia, thorax and antenna

8. Female genitalia in lateral view. MP-median prong of IXth tergite, PTGR- procti-
   ger, 2VLF-second valvifer, VP-ventral prong of IXth tergite, IVLF-first valvifer,
   AP-anterior prong of 2nd valvula, 3VL-third valvula, IVL-first valvula, 2VL-second
   valvula.
9. Lateral view thorax, NOT-notaulices, MSCT-mesoscutum, TG-tegula, SCTL-
   scutellum, AX-axilla, MTN-metanotum, PSCTL-postscutellum, PROP-propodeum, SP-spiracle, T_1-tergite I, PN-pronotum, PP-propleuron, CX_1-first coxa,
   MSPL-mesopleuron, MTPL-metapleuron, CX_2-second coxa, CX_3-third coxa.
10. Ventral view of male genitalia. P-penis, VP-valvae of penis, D-digitus, C-cuspis,
    GF-gonoforceps, APD-apodeme; GR-gonostipital rami, GX-gonocoxite, GB-
    gonobase.
11. Antenna (basal part). S-scape, P-pedicel, F-funicle.

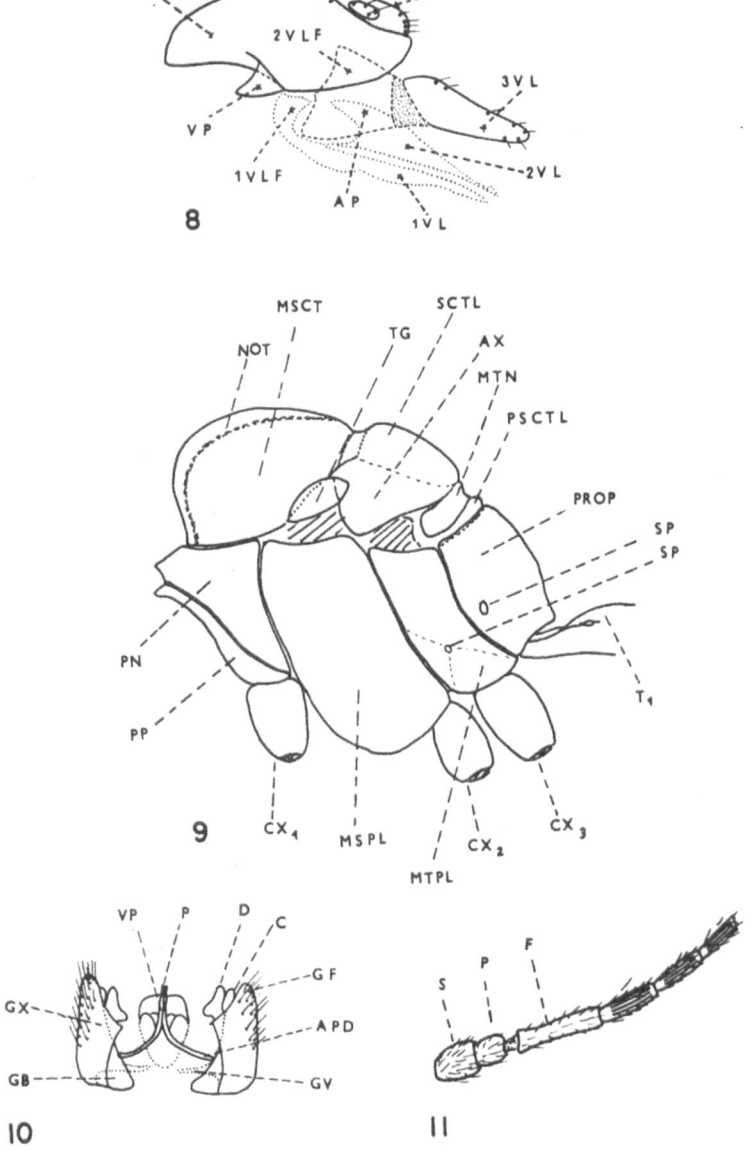

Plate IV. Female genitalia of *Trioxys* species

12. *Trioxys (Trioxys) asiaticus* TELENGA
13. *Trioxys (Binodoxys) brunnescens* STARÝ & SCHLINGER
14. *Trioxys (Binodoxys) indicus* SUBBA RAO & SHARMA
15. *Trioxys (Fissicaudus) confucius* MACKAUER
16. *Trioxys (Binodoxys) communis* GAHAN
17. *Trioxys (Binodoxys) carinatus* STARÝ & SCHLINGER

153

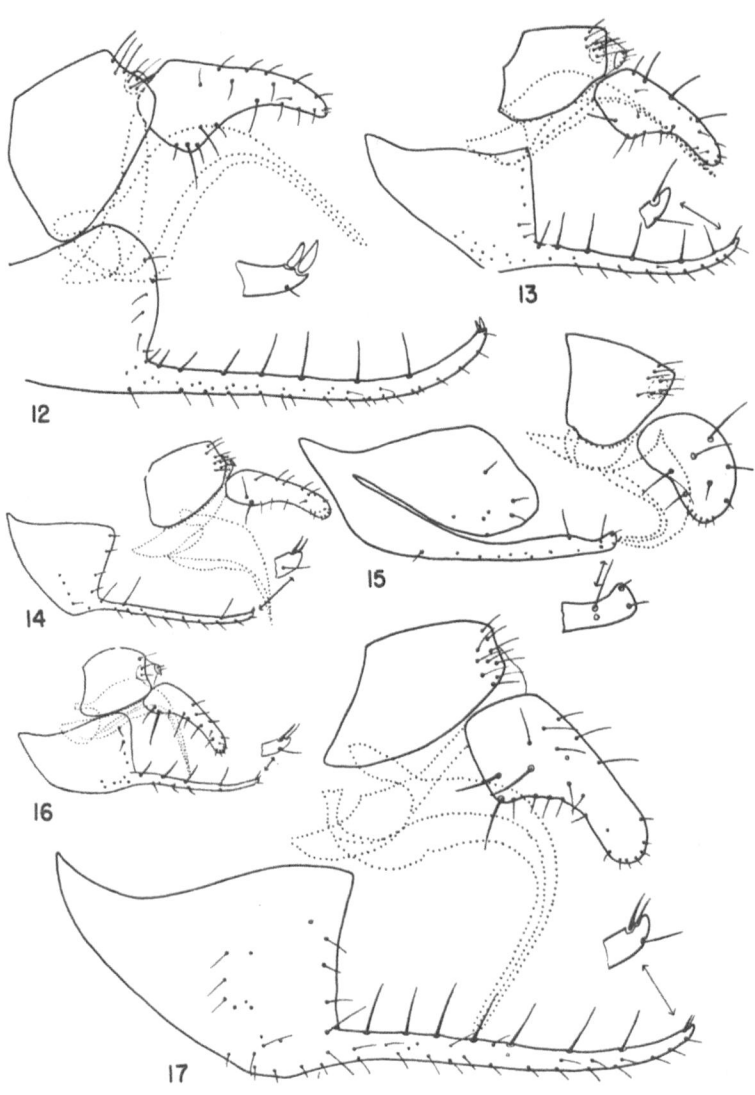

Plate V. Female genitalia of *Trioxys* species

18. *Trioxys (Binodoxys)* "rietscheli" group
19. *Trioxys (Binodoxys) sinensis* MACKAUER
20. *Trioxys (Trioxys) luteolus* STARÝ & SCHLINGER
21. *Trioxys (Binodoxys) orientalis* STARÝ & SCHLINGER

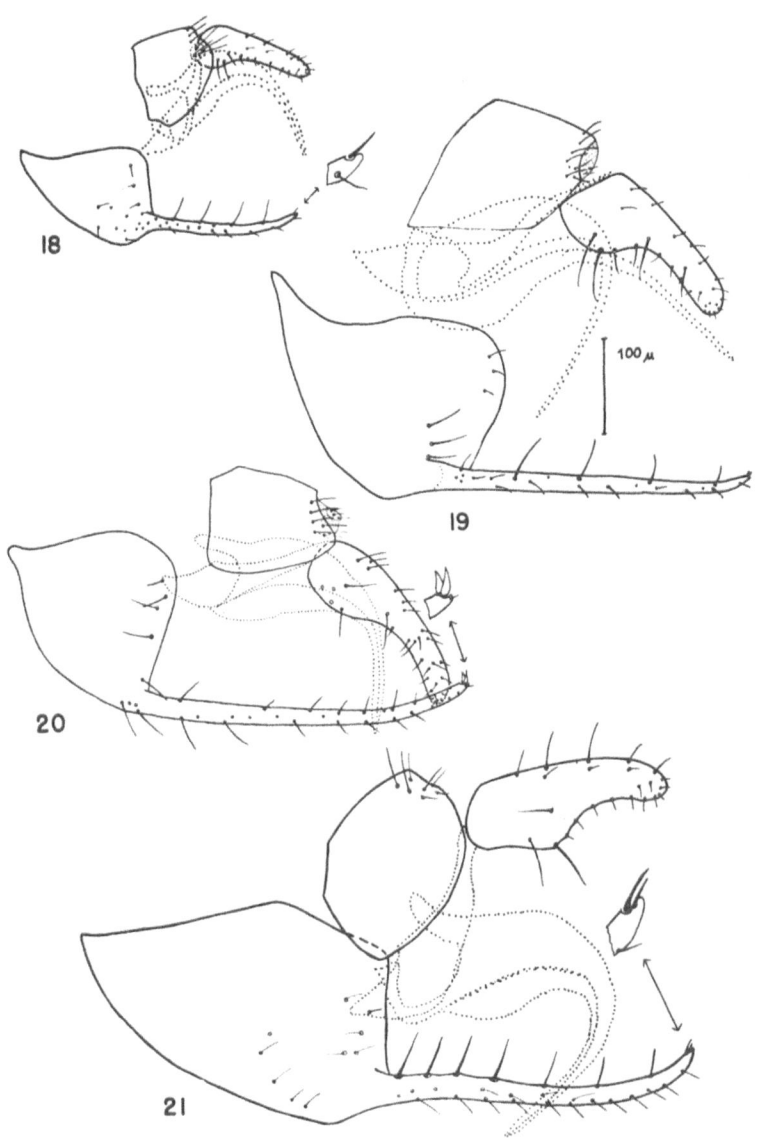

18

19

100 μ

20

21

Plate VI. Female genitalia of various Aphidiidae

22. *Aphidius "gifuensis"* group
23. *Aphidius salignae* WATANABE
24. *Aphidius absinthii* MARSHALL
25. *Aphidius salicis* HALIDAY
26. *Aphidius areolatus* ASHMEAD
27. *Diaeretiella rapae* (M'INTOSH)
28. *Protaphidius wissmannii* (RATZEBURG) [from GOIDANICH, 1934]
29. *Diaeretus leucopterus* (HALIDAY)
30. *Monoctonus (Monoctonus) woodwardiae* STARÝ & SCHLINGER
31. *Monoctonus (Monoctonus) similis* STARÝ & SCHLINGER

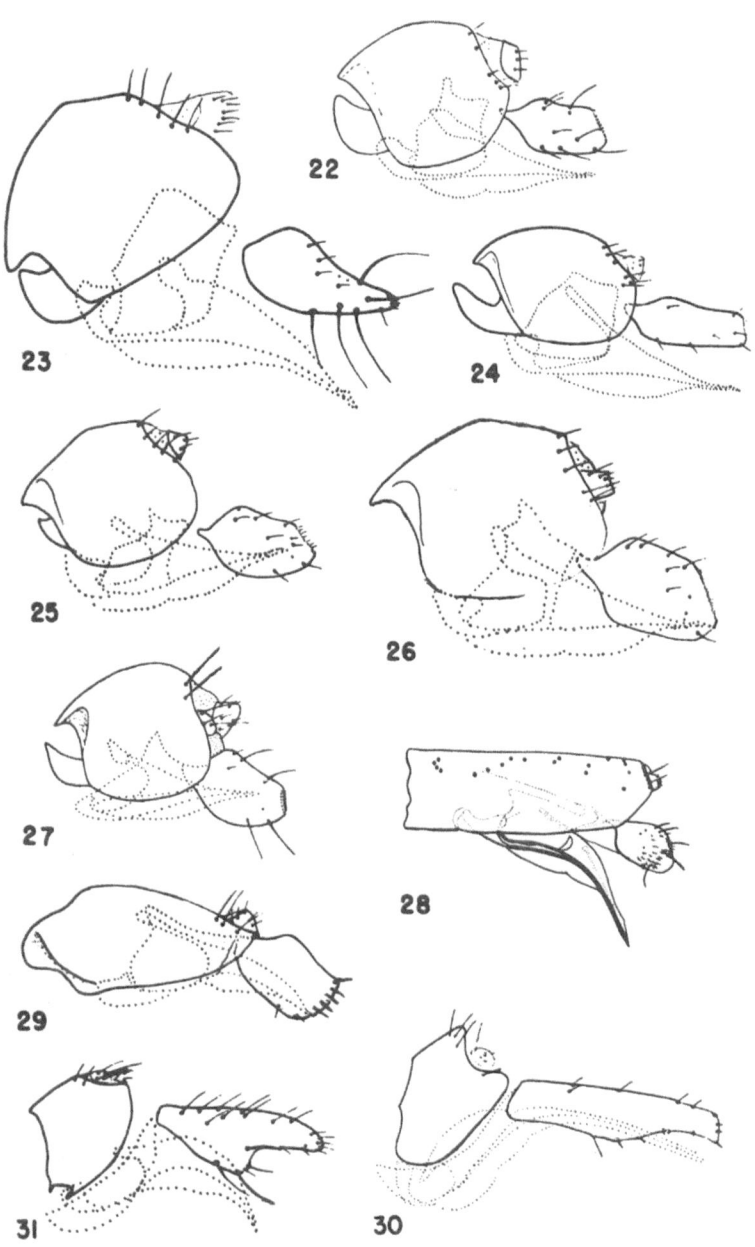

Plate VII. Female genitalia of *Ephedrus* species

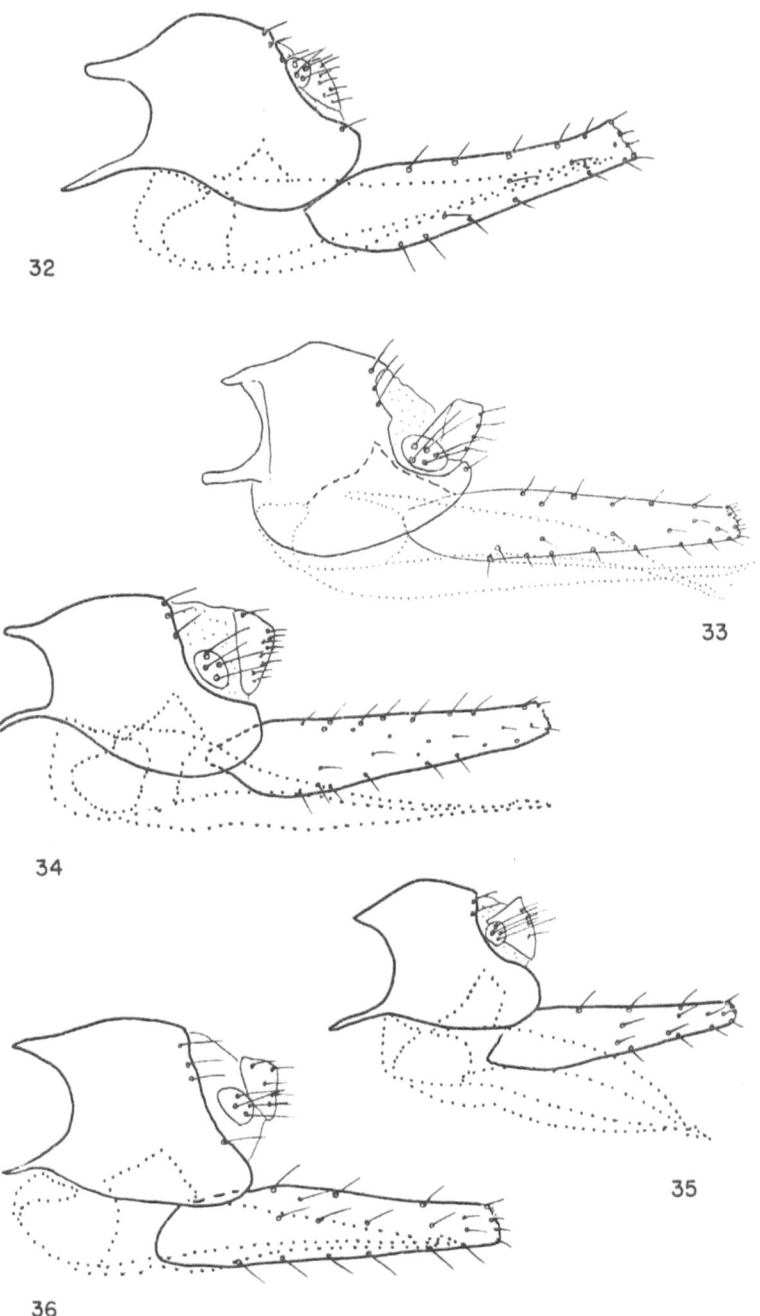

32

33

34

35

36

Plate VIII. Female genitalia of various Aphidiidae

37. *Lysiphlebia japonica* (ASHMEAD)
38. *Lipolexis scutellaris* MACKAUER
39. *Lysiphlebus salicaphis* (FITCH)
40. *Lipolexis gracilis* FÖRSTER
41. *Lysiphlebus "delhiensis"* group.
42. *Bioxys japonicus* STARÝ & SCHLINGER

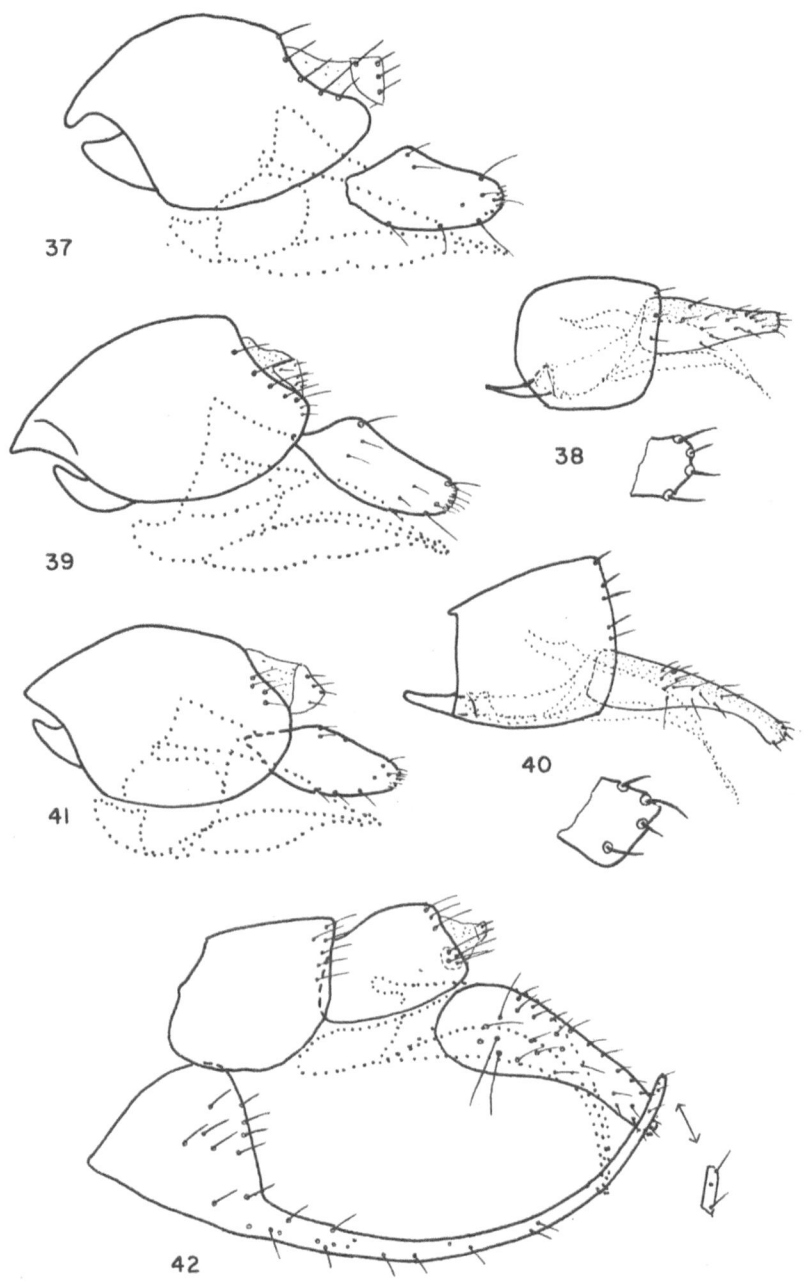

37

38

39

40

41

42

Plate IX. Female genitalia of various Aphidiidae

43. *Ephedrus (Lysephedrus) validus* HALIDAY
44. *Praon orientale* STARÝ & SCHLINGER
45. *Pauesia unilachni* (GAHAN)
46. *Lipolexis oregmae* (GAHAN)
47. *Praon glabrum* STARÝ & SCHLINGER

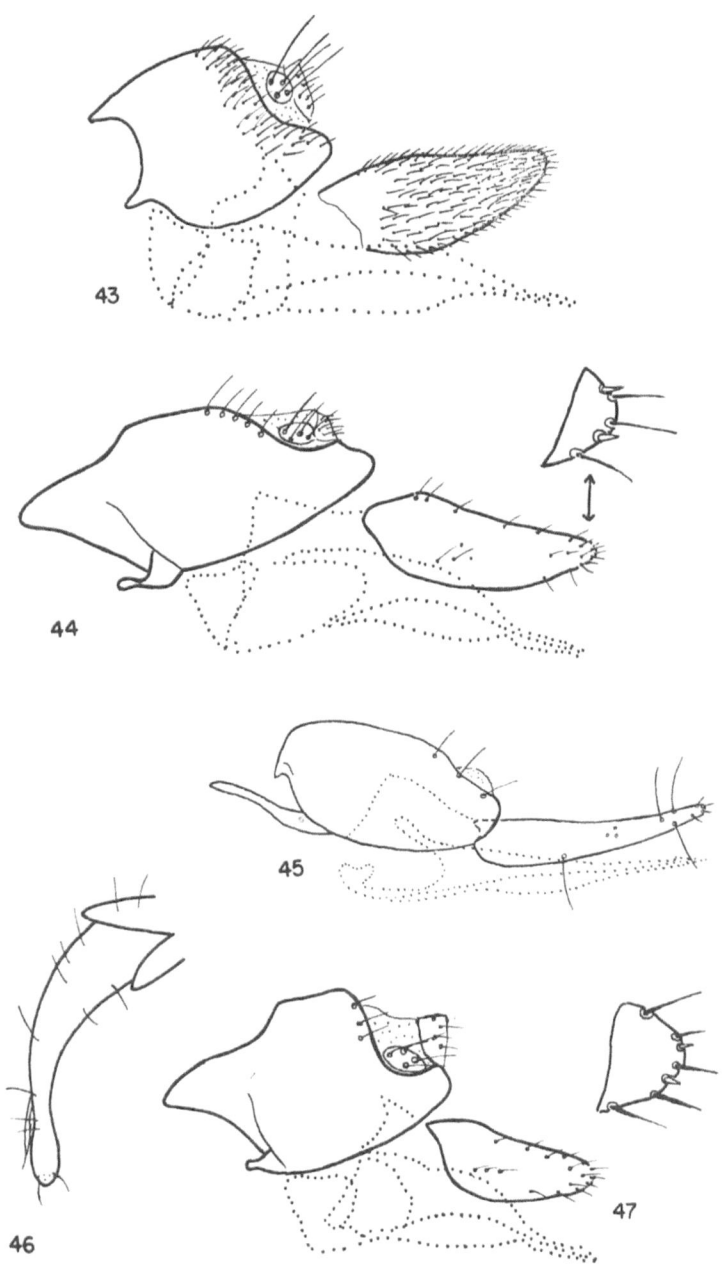

Plate X. Female genitalia of *Pauesia* and *Trioxys* sp.

48. *Pauesia infulata* (HALIDAY)
49. *Pauesia pini* (HALIDAY)
50. *Trioxys (Binodoxys) struma* GAHAN

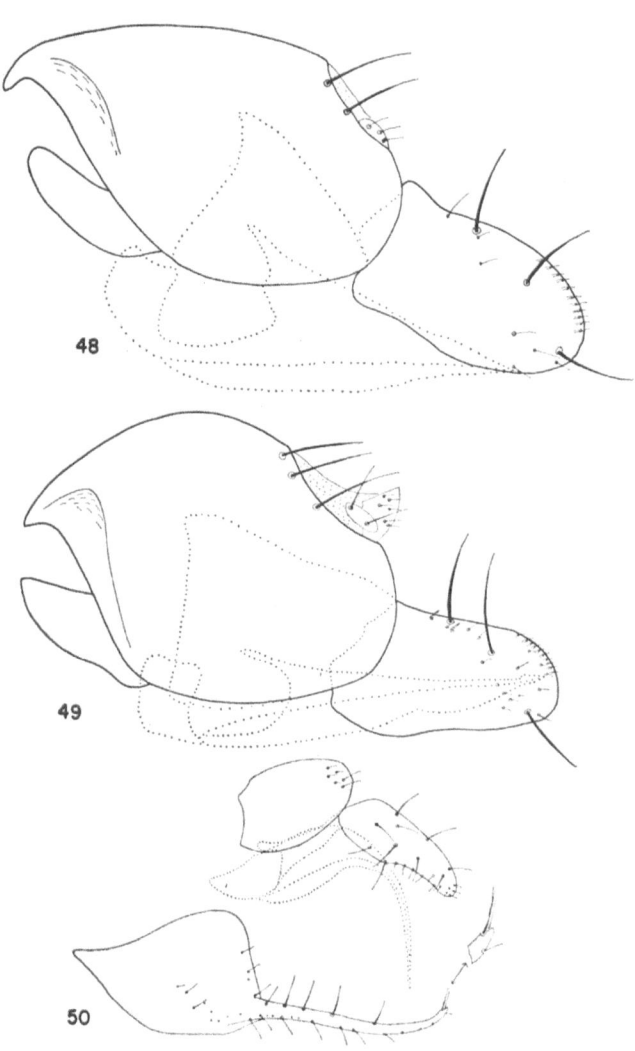

48

49

50

Plate XI.

51. Head of *Pauesia infulata* (HALIDAY), ♀
52. Propodeum of *Pauesia tropicalis* STARÝ & SCHLINGER, ♀
53. Propodeum of *Protaphidius wissmannii* (RATZEBURG), ♀
54. Propodeum of *Pauesia pini* (HALIDAY), ♀
55. Propodeum of *Pauesia infulata* (HALIDAY), ♀

167

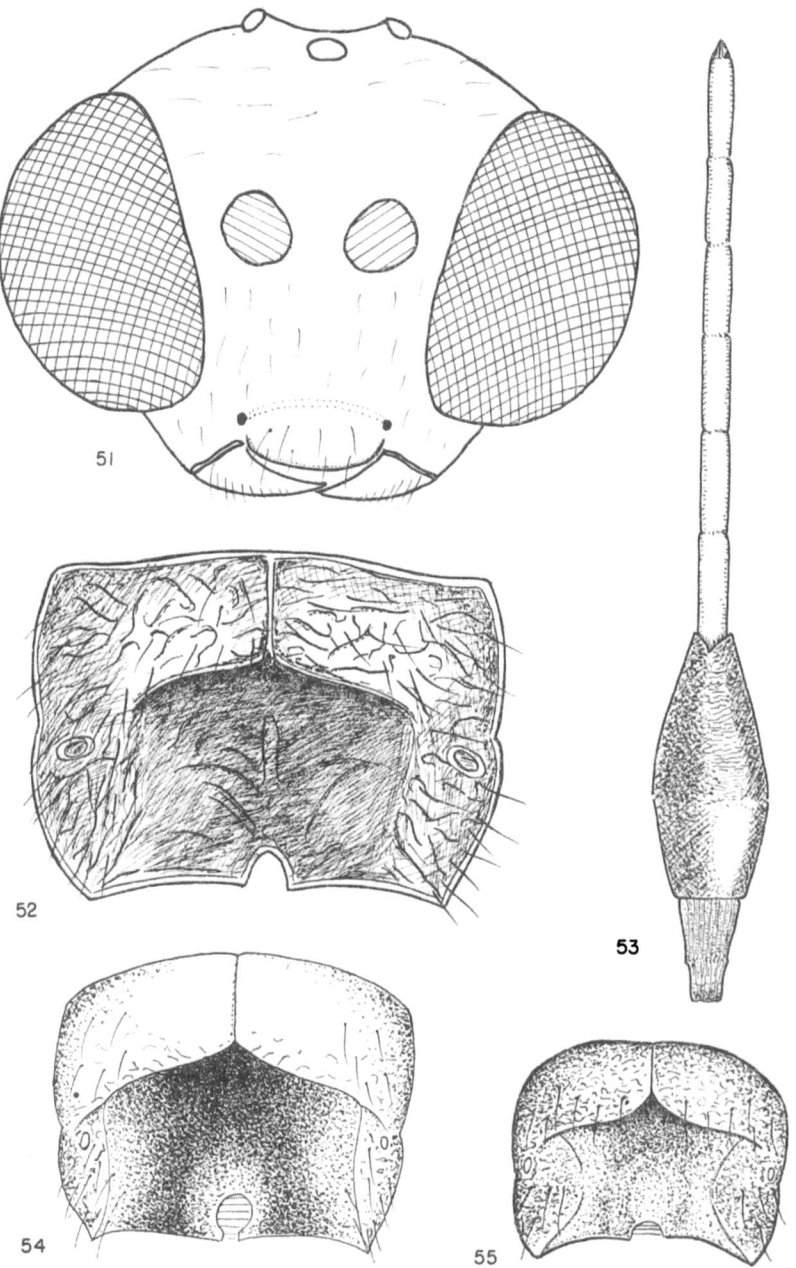

51

52

53

54

55

Plate XII.

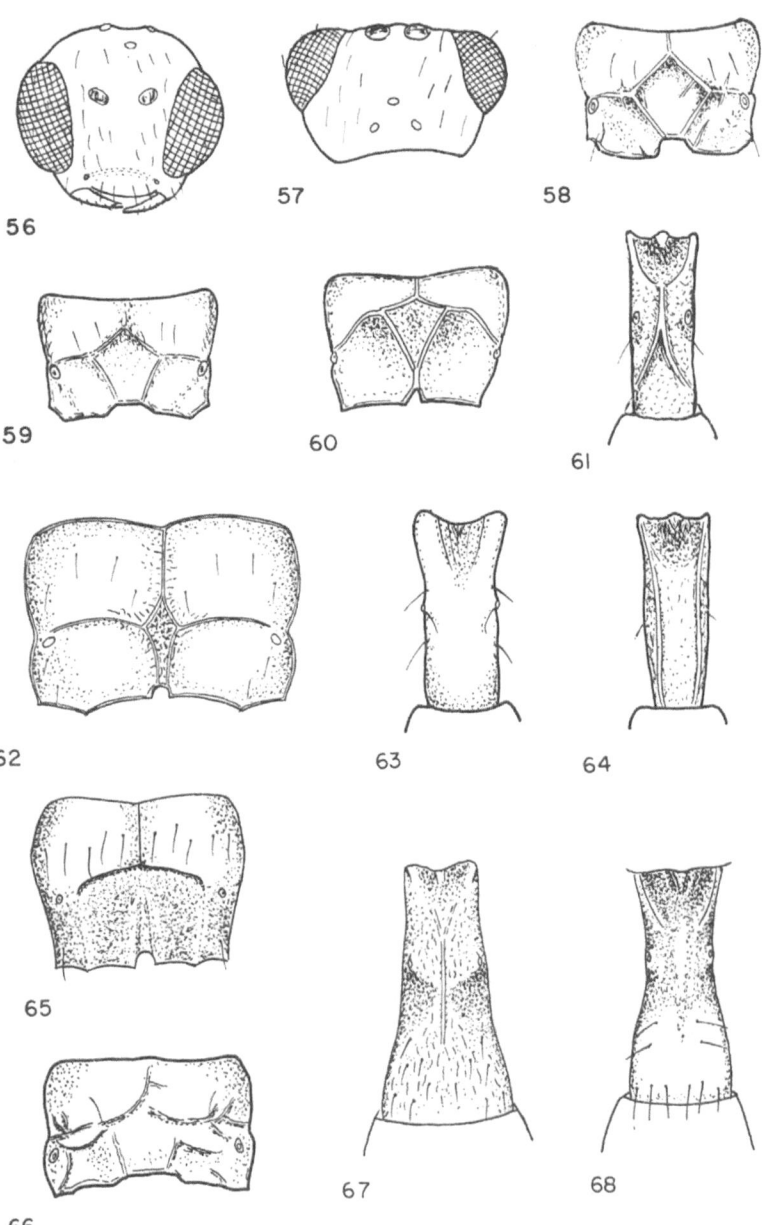

56

57

58

59

60

61

62

63

64

65

66

67

68

Plate XIII.

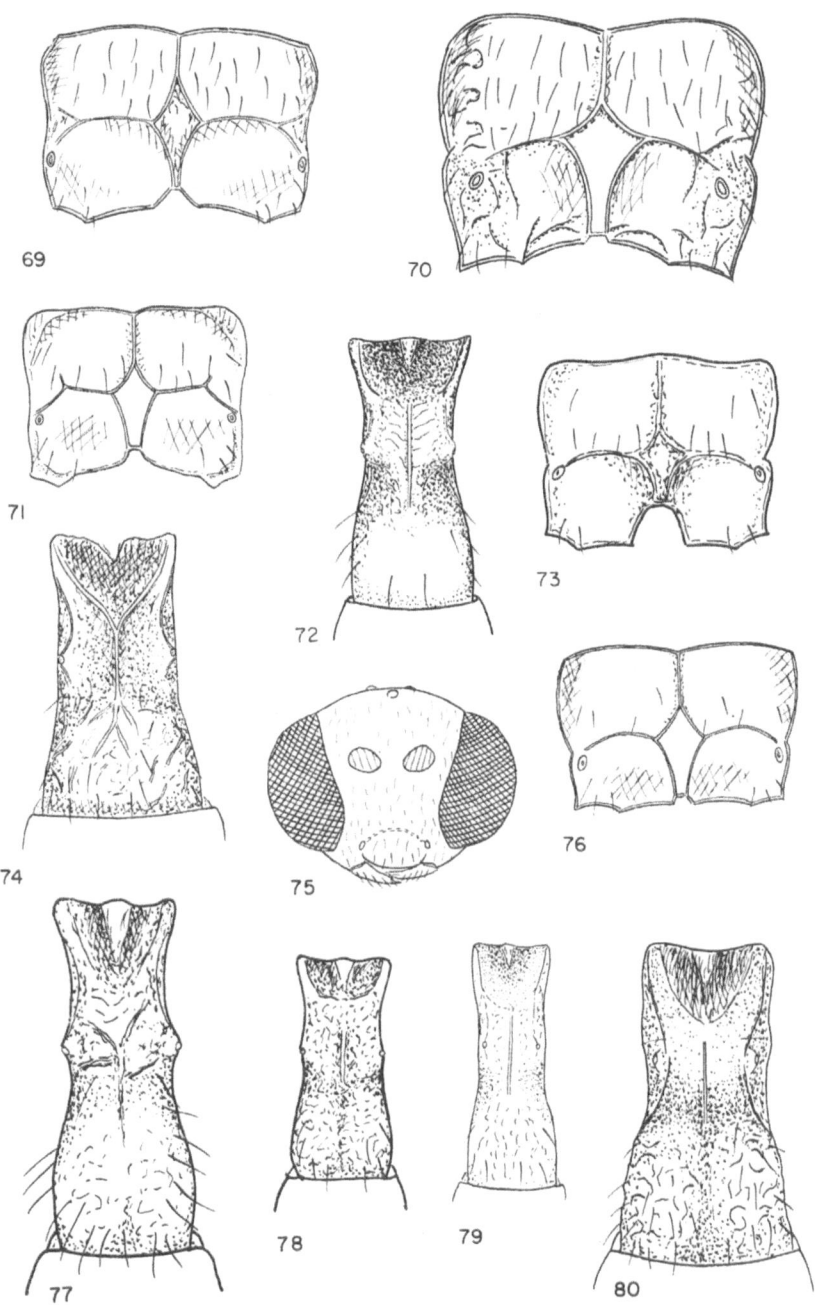

69

70

71

72

73

74

75

76

77

78

79

80

Plate XIV.

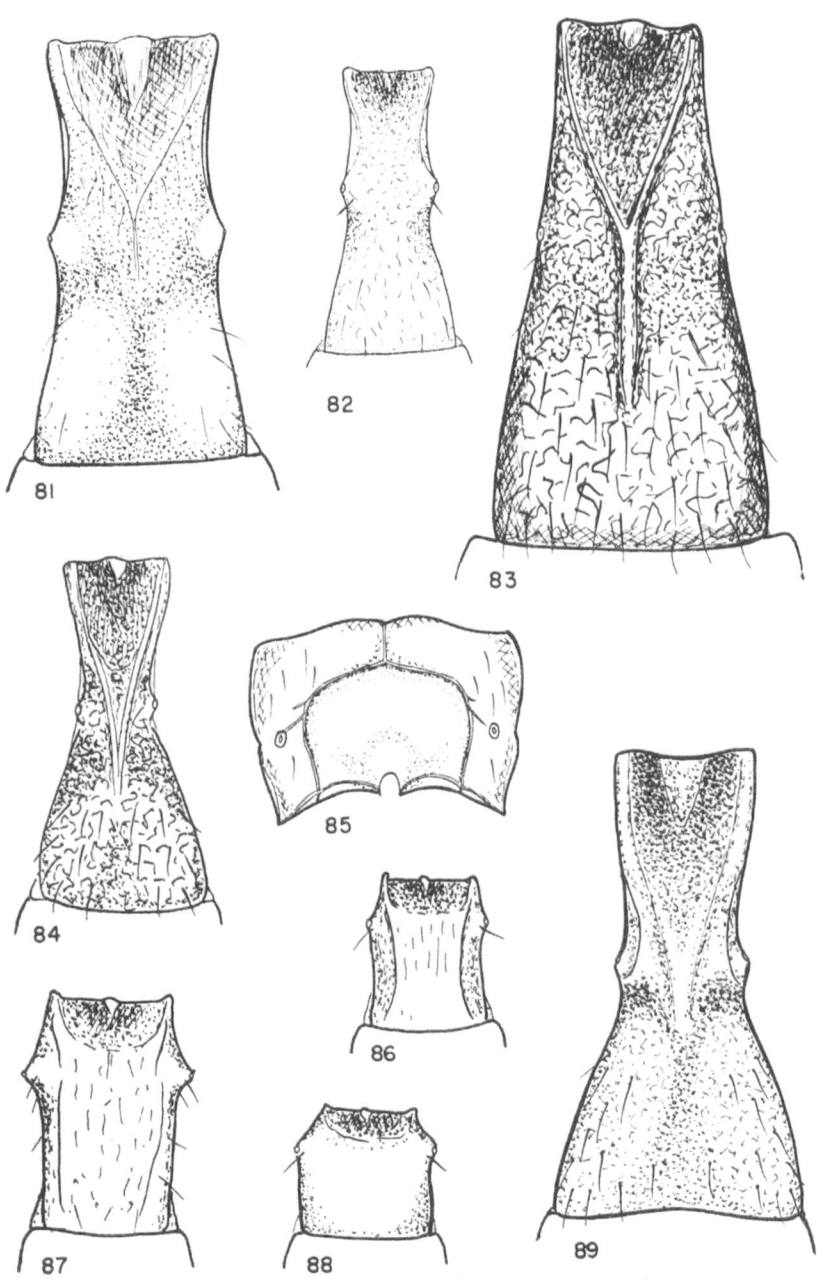

Plate XV.

175

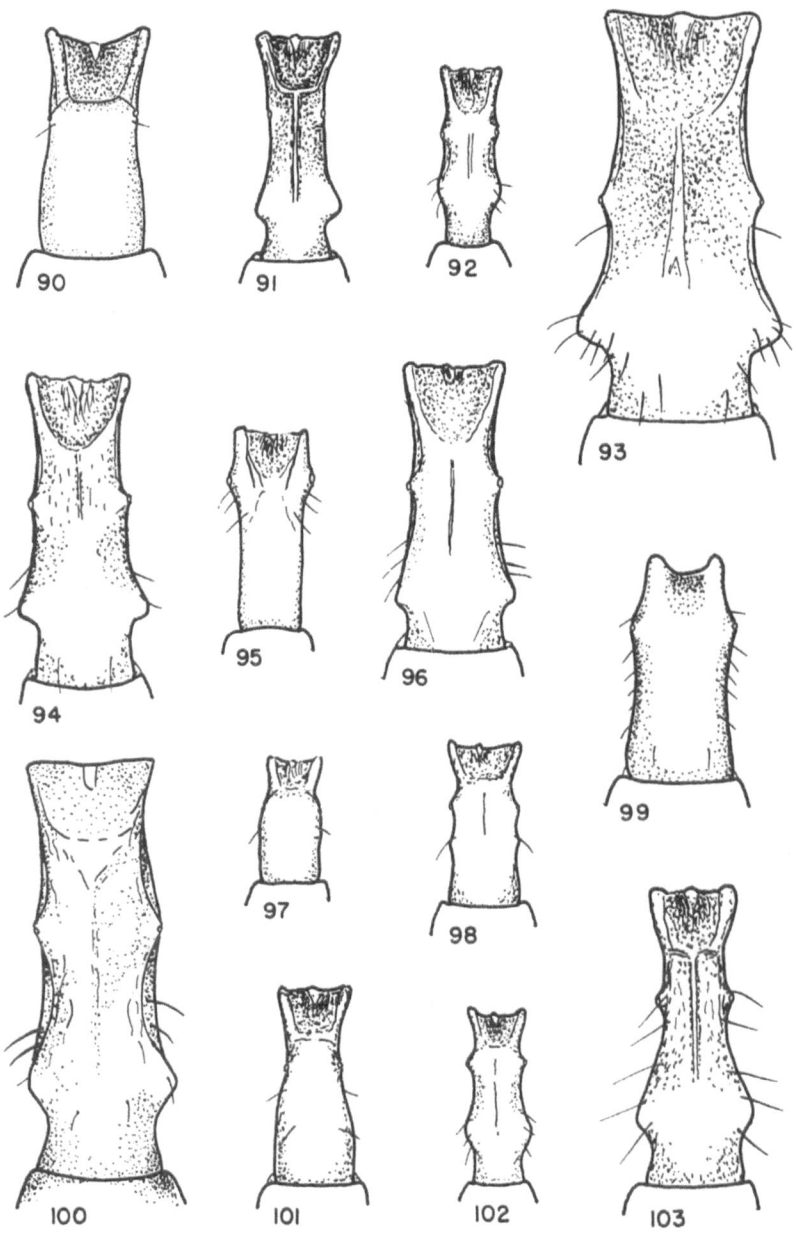

Plate XVI.

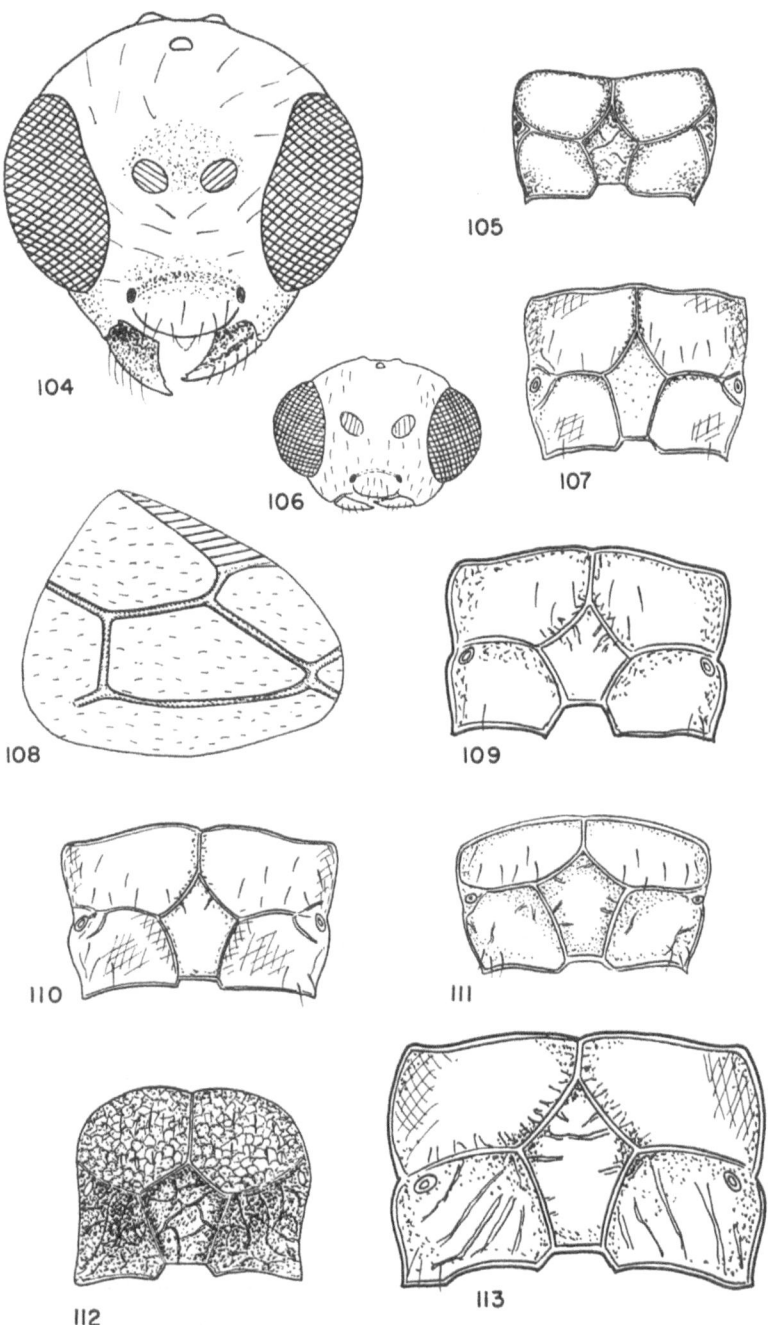

Plate XVII.

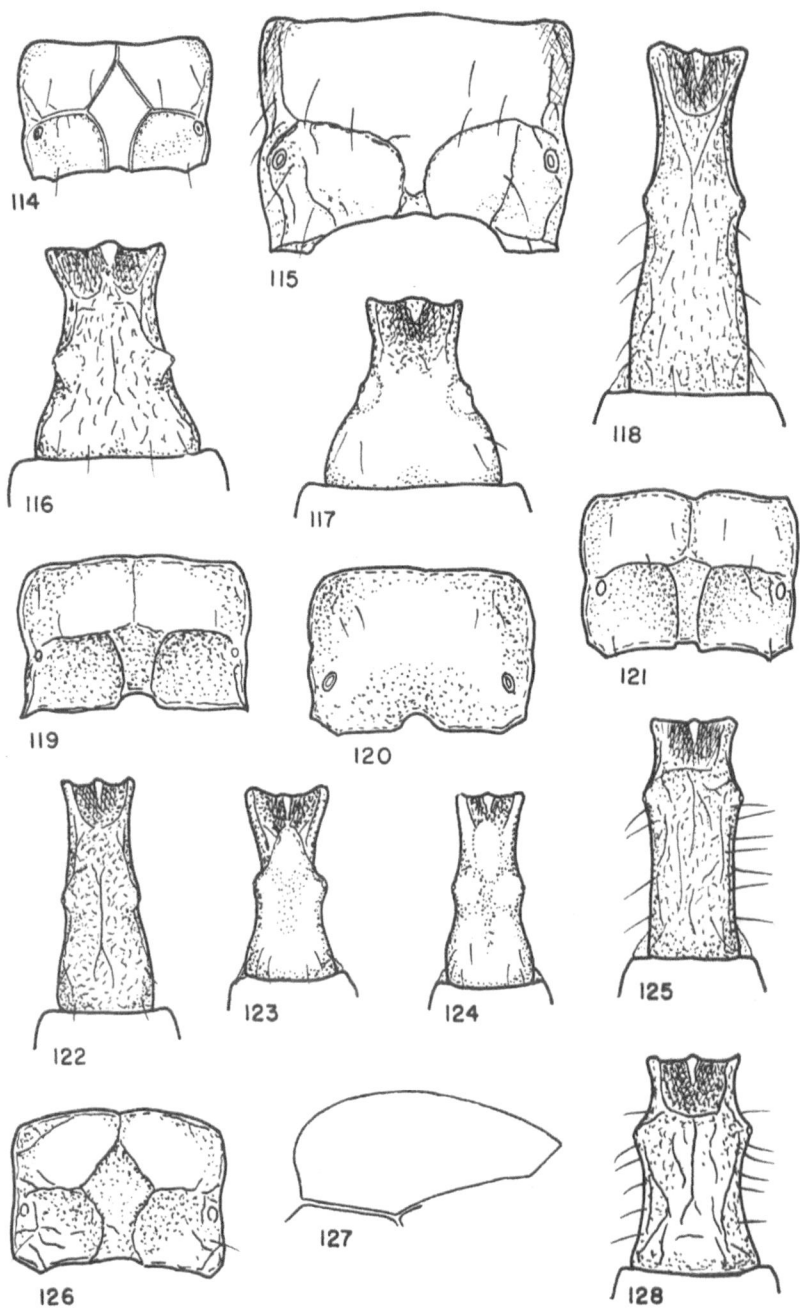

Plate XVIII. Propodea of *Trioxys* species

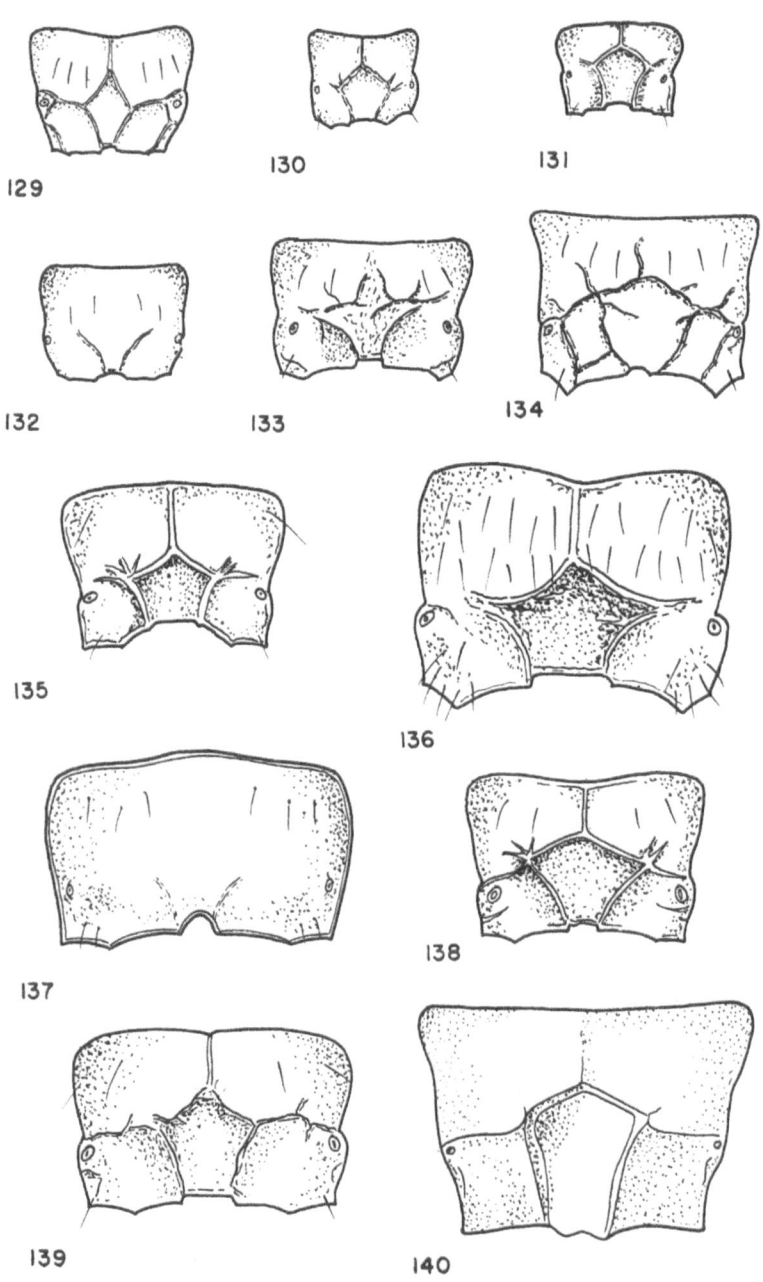

129

130

131

132

133

134

135

136

137

138

139

140

182

Plate XIX.

141. Tergite I of *Ephedrus (Ephedrus) persicae* FROGGATT, ♀
142. Tergite I of *Ephedrus (Ephedrus) plagiator* (NEES), ♀
143. Tergite I of *Ephedrus (Lysephedrus) validus* HALIDAY, ♀
144. Tergite I of *Ephedrus (Ephedrus) orientalis* STARÝ & SCHLINGER, ♀
145. Tergite I of *Ephedrus (Ephedrus) lacertosus* (HALIDAY), ♀
146. Tergite I of *Ephedrus (Ephedrus) plagiator* (NEES) ♂ [from paratype of *E. japonicus* Ash.]
147. Tergite I of *Ephedrus (Ephedrus) campestris* STARÝ, ♀
148. Antennal segments $F_1 + F_2$ of *Ephedrus (Ephedrus) lacertosus* HALIDAY, ♀
149. Antennal segments $F_1 + F_2$ of *Ephedrus (Ephedrus) plagiator* (NEES), ♀
150. Antennal segments $F_1 + F_2$ of *Ephedrus (Ephedrus) campestris* STARÝ, ♀
151. Antennal segments $F_1 + F_2$ of *Ephedrus (Ephedrus) orientalis* STARÝ & SCHLINGER ♀

183

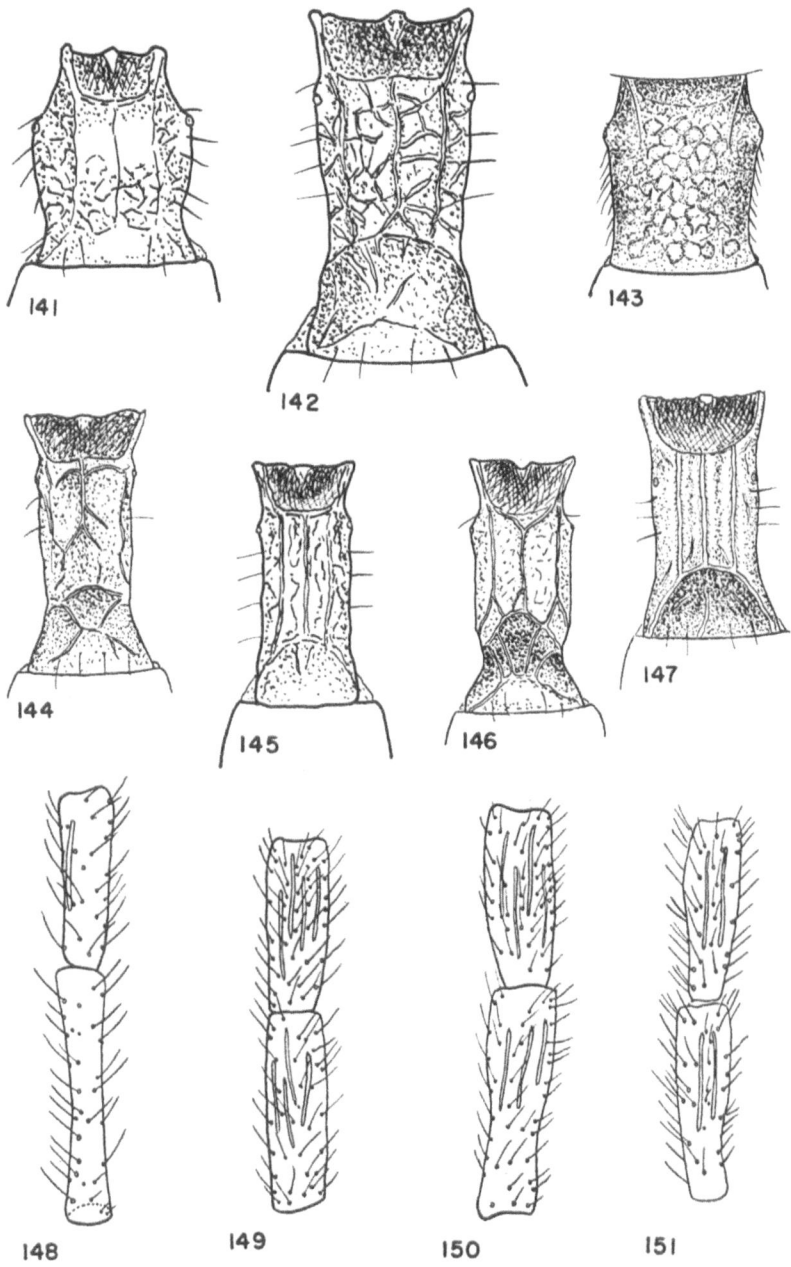

Plate **XX**.

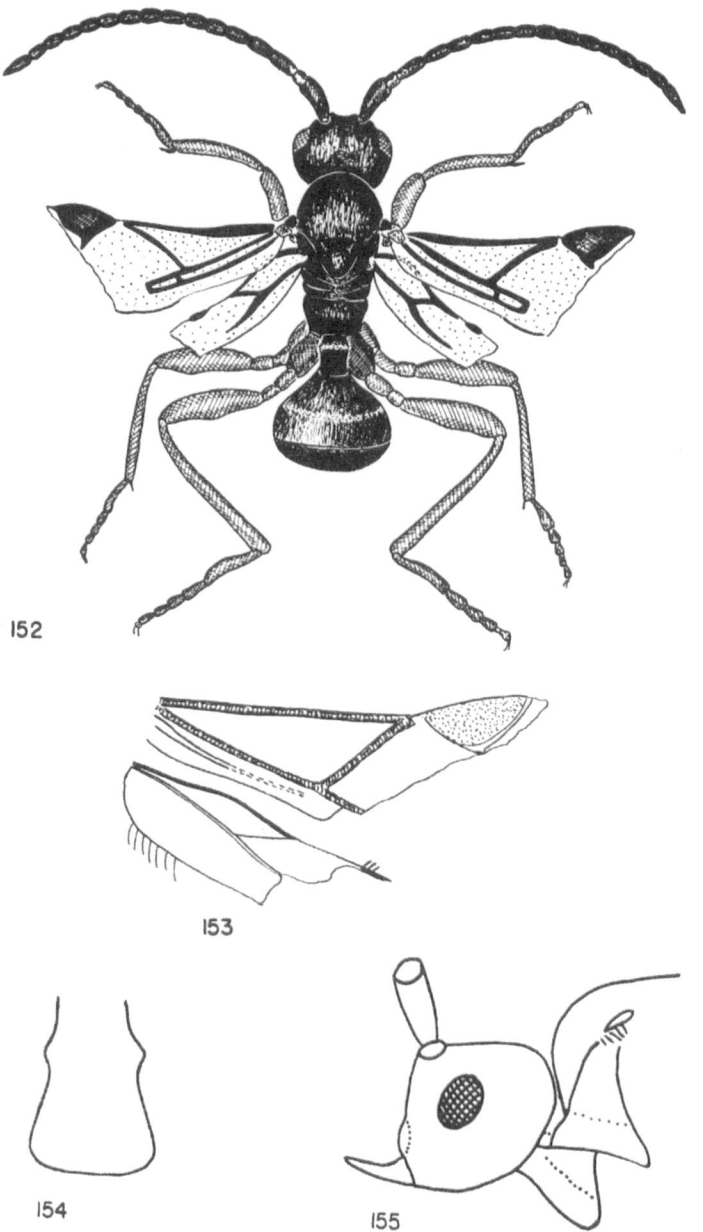

152

153

154   155

Plate XXI. Wings

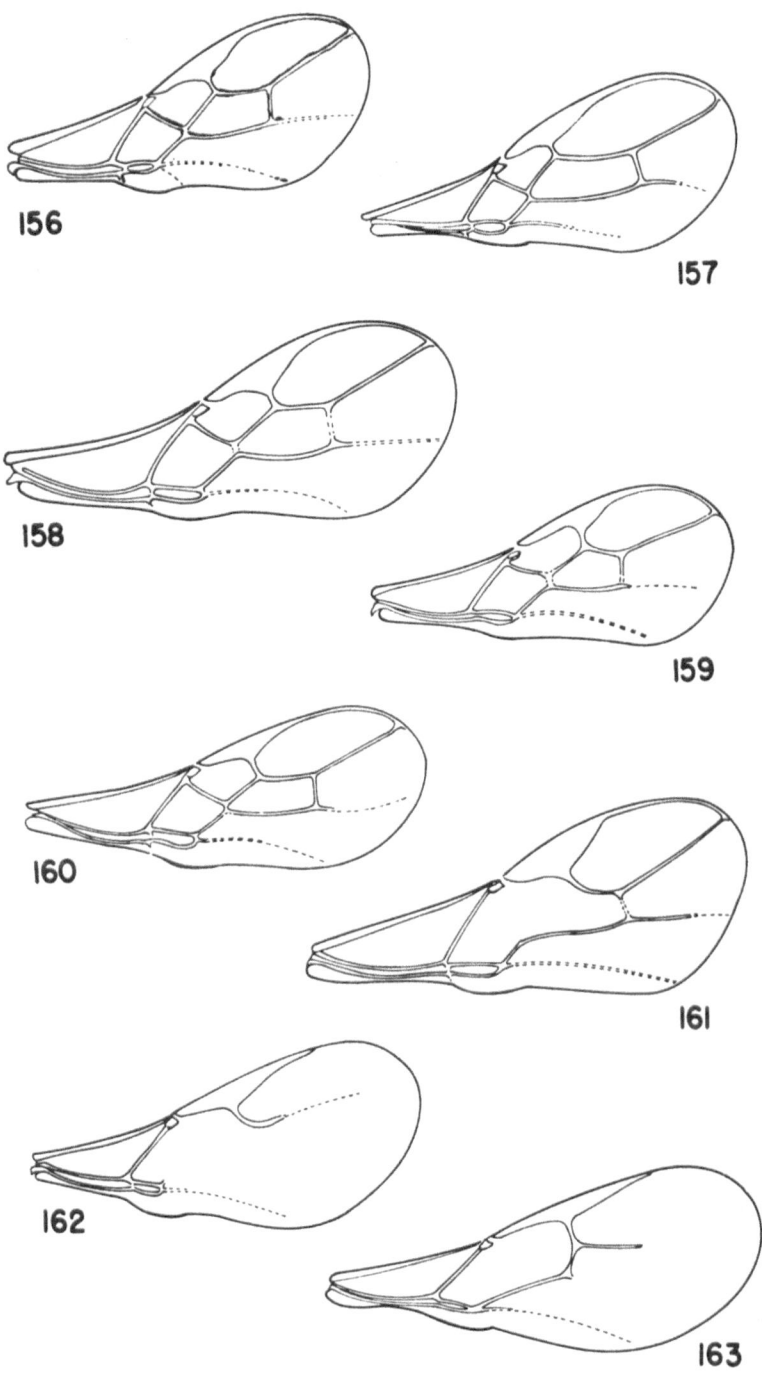

Plate XXII. Wings

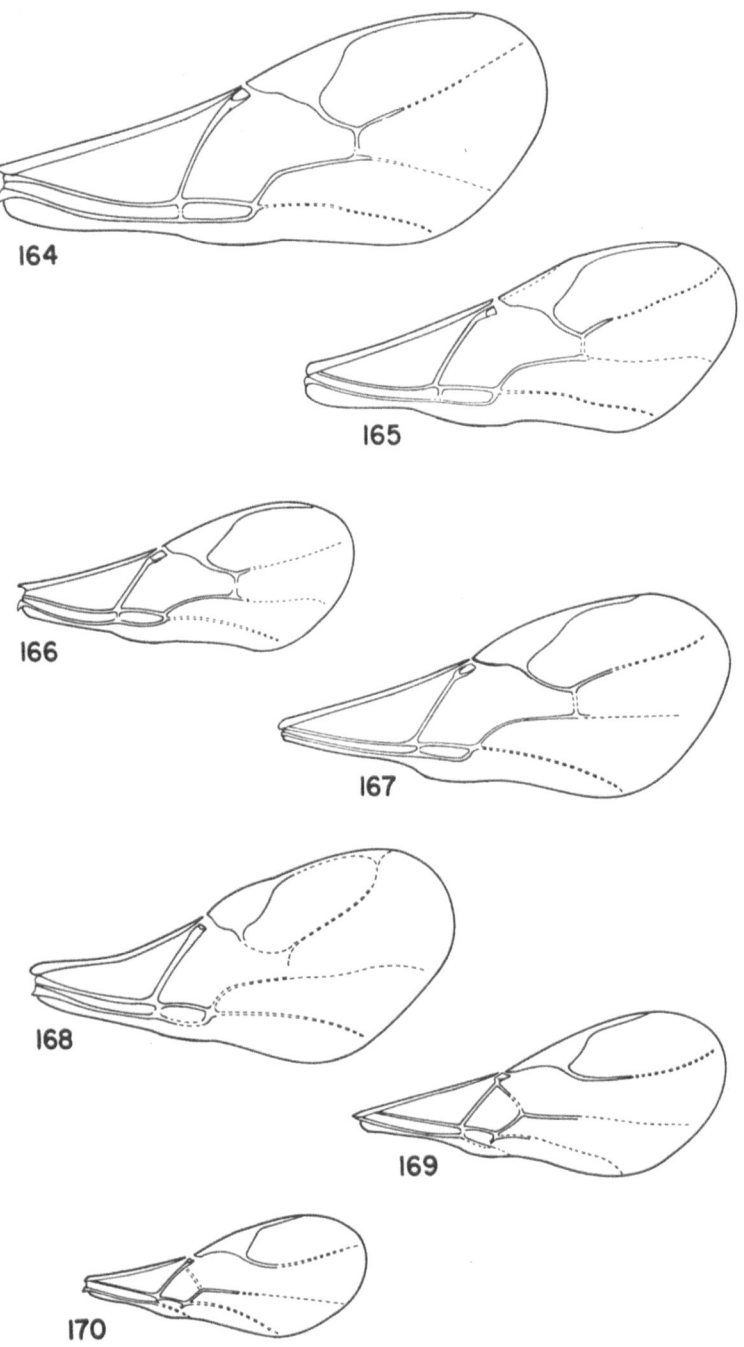

189

Plate XXIII. Wings

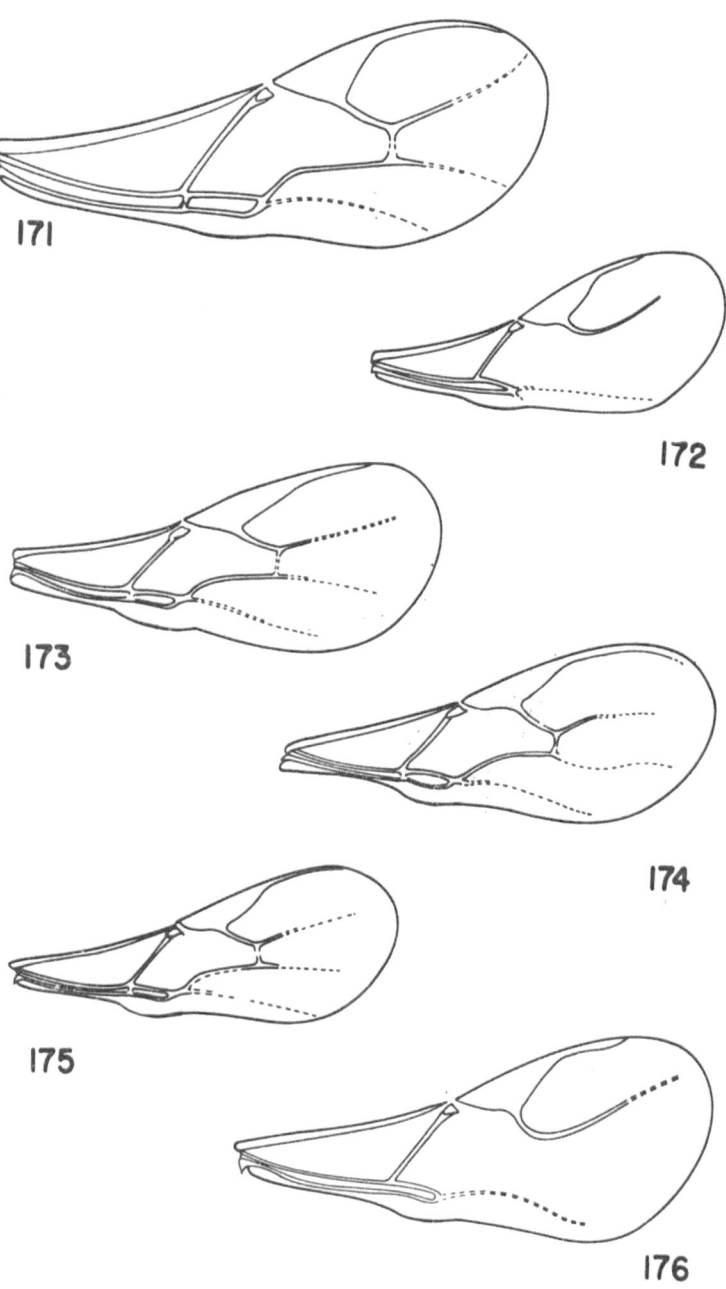

171

172

173

174

175

176

Plate XXIV. Wings

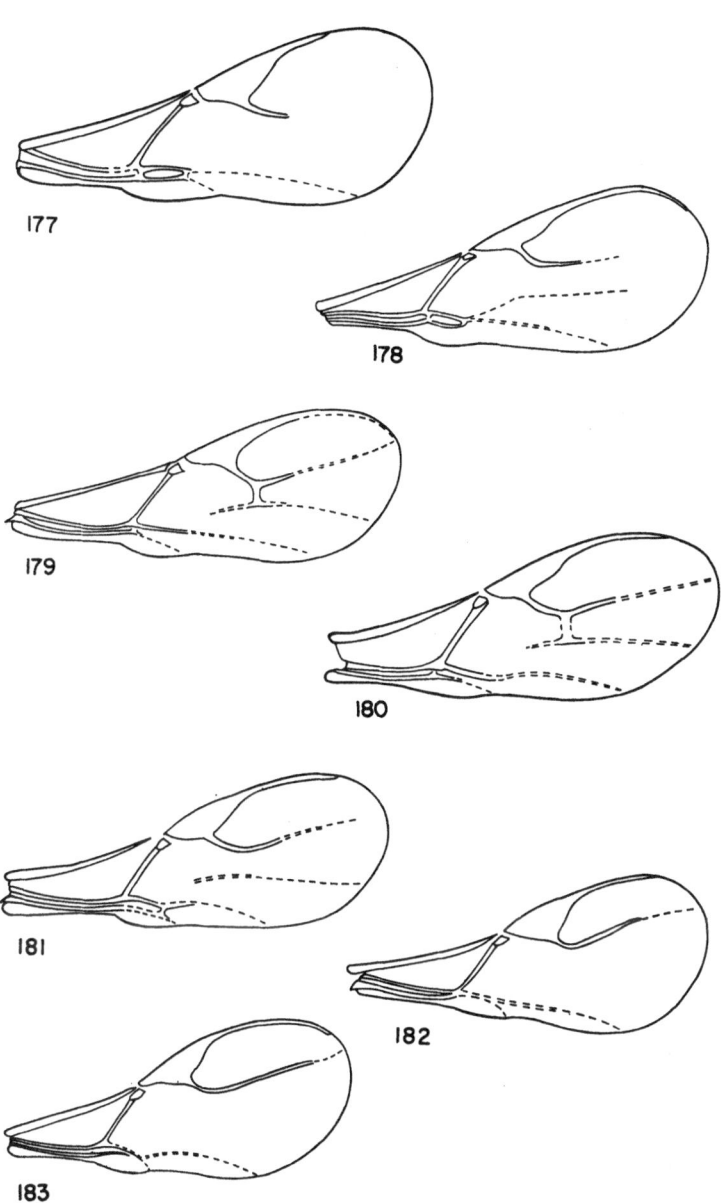

Plate XXV. Wings

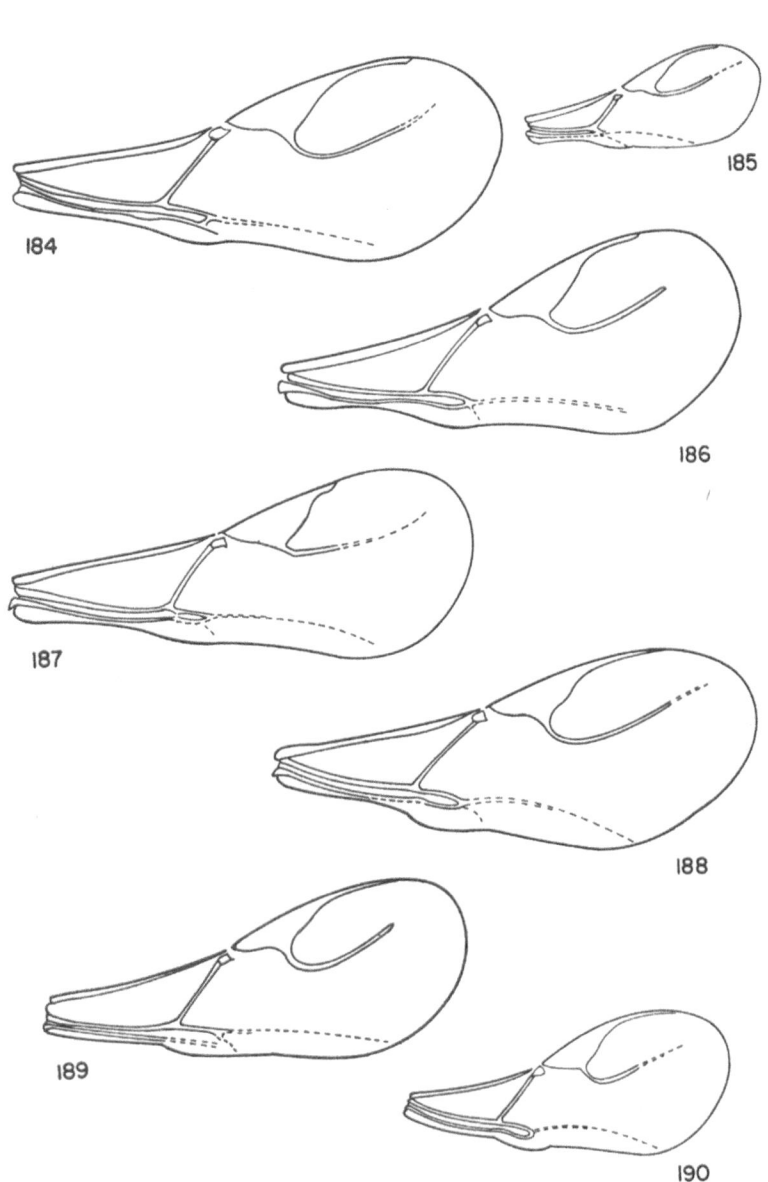

SUBJECT INDEX

199

PLANT INDEX

Abies alba, 84
  sachalinensis, 93, 97
Acer palmatum 17
  spp., 5, 17, 127
Achillea millefolium, 15, 44
  nobilis, 44
  sudetica, 15
Aethionema saxatile, 36
Alliaria officinalis, 37
Angelica silvestris, 22
Anthemis tinctoria, 15
  sp., 50
Anthriscus silvestris, 21
Apium graveolens, 21
  sp., 62
Aralia spp., 59, 102
Archangelica officinalis, 58
Arctium lappa, 65
Artemisia absinthium, 14, 15, 44
  campestris, 15
  manischmidtiana, 15, 45
  sp., 14
Bambusa sp., 28
Berberis vulgaris, 59
Beta vulgaris, 38, 65
Betula sp., 36, 100
Borago officinalis, 58
Brassica napus 37
  oleracea, 36, 37
  oleracea var. capitata, 38
  sp., 36, 51
Bupleurum falcatum, 65
Callicarpa japonica, 58
Caragana arborescens, 58
Carduus nutans, 45
  sp., 65
Centaurea cyanus, 65
  scabiosa, 45
  stoebe, 15, 45
Chaenomelles japonica, 51
  sinensis, 58
Chenopodium album, 36, 37
  vulvaria, 21
Chrysanthemum sp., 45, 50, 104
Cichorium intybus, 44, 65
Cirsium sp., 65

Citrus aurantium ssp. maxima, 68
  sinensis, 19, 50, 71
  sp., 103, 119
Convolvulus sp., 102
Cornus sanguinea, 64
Corydalis platycarpa, 19, 29
Corylus avellana, 59
Crataegus oxyacantha, 51
  pinnatigida, 71, 102
Crepis biennis, 44
Crithmum maritimum, 21
Daucus carota, 21
Delphinium sp., 19
Deutzia crenata, 60, 103
Echium vulgare, 65
Elæagnus sp. 50, 59
Epipactis latifolia, 58
Ervum hirsutum, 46
Euonymus europæa, 58
Euphorbia cyparissias, 65
Euphrasia inumai, 19
Festuca pratensis, 60
Ficus sp., 31, 33, 90, 120, 123
Foeniculum vulgare, 37, 50, 71
Fraxinus excelsior, 59
Gossypium hirsutum, 125, 126
Hibiscus sp., 50, 65, 114, 115
Hieracium sp., 44
Hordeum distychum, 60
Humulus lupulus, 51
Impatiens roylei, 58
Iris variegata, 65
Juniperus communis, 84
Keteleeria davidiana, 127
Koelreuteria, 5, 17
Lactuca quercina, 44
  sativa, 59, 102
Larix sp. 88
Leontodon hispidus, 44
Ligustrum vulgare, 51
Lolium sp., 38
Lonicera morrowii, 58
  sp., 58
Malcolmia maritima, 38
Malus silvestris, 50, 51, 59
Malva sp., 19